Hot X

ALGEBRA EXPOSED

Also by Danica McKellar

*Math Doesn't Suck: How to Survive Middle School Math
Without Losing Your Mind or Breaking a Nail*

Kiss My Math: Showing Pre-Algebra Who's Boss

HOT X

ALGEBRA EXPOSED

Danica McKellar

HUDSON
STREET
PRESS

HUDSON STREET PRESS
Published by the Penguin Group
Penguin Group (USA) Inc., 375 Hudson Street, New York, New York 10014, U.S.A.
Penguin Group (Canada), 90 Eglinton Avenue East, Suite 700, Toronto, Ontario, Canada M4P
2Y3 (a division of Pearson Penguin Canada Inc.)
Penguin Books Ltd., 80 Strand, London WC2R 0RL, England
Penguin Ireland, 25 St. Stephen's Green, Dublin 2, Ireland (a division of Penguin Books Ltd.)
Penguin Group (Australia), 250 Camberwell Road, Camberwell, Victoria 3124, Australia (a
division of Pearson Australia Group Pty. Ltd.)
Penguin Books India Pvt. Ltd., 11 Community Centre, Panchsheel Park, New Delhi – 110 017,
India
Penguin Group (NZ), 67 Apollo Drive, Rosedale, North Shore 0632, New Zealand (a division of
Pearson New Zealand Ltd.)
Penguin Books (South Africa) (Pty.) Ltd., 24 Sturdee Avenue, Rosebank, Johannesburg 2196,
South Africa

Penguin Books Ltd., Registered Offices: 80 Strand, London WC2R 0RL, England

First published by Hudson Street Press, a member of Penguin Group (USA) Inc.

First Printing, August 2010
10 9 8 7 6 5 4 3 2 1

PHOTO CREDITS: Page 28, © Henry Fellows; Page 162, © Eric Parry; Page 196, © Katinka
Rodriguez; Page 222, © Cathy Ann Tudor; Page 306, © Jaimee Young; Page 320, © Kimberlee
West; Page 383, © Michael Madland.

REGISTERED TRADEMARK—MARCA REGISTRADA

HUDSON
STREET
PRESS

LIBRARY OF CONGRESS CATALOGING-IN-PUBLICATION DATA

McKellar, Danica.
   Hot X : algebra exposed / Danica McKellar.
      p.   cm.
   Includes index.
   ISBN 978-1-59463-070-5 (hardcover : alk. paper)   1. Algebra—Study and teaching (Middle
school)   2. Algebra—Study and teaching (Secondary)   3. Algebra—Popular works.   I. Title.
   QA159.M345 2010
   512—dc22

                                                                          2010018163

Printed in the United States of America
Set in ITC Stone Informal

This book is printed on acid-free paper. ∞

$\mathcal{I}$ dedicate this book to YOU! Let's build the next generation of kick-ass gals who look algebra in the eye and say, "I shall be your master."

Acknowledgments

Thank you to my parents, Mahaila and Chris, for always encouraging and believing in me! Thank you to my best friend in the whole world, my sister Crystal, and to the rest of my wonderful family, including (but not limited to) Chris Jr., Connor, Grammy, Opa, Lorna (and kids!), Jimmy, Molly, and now the Vertas! You all mean so much to me.

Thank you to the team at HSP—Clare Ferraro, Liz Keenan, and the new dynamic duo, Caroline Sutton and Meghan Stevenson! And to Luke Dempsey—you and your birds will be missed. Thank you to my wonderful agent Laura Nolan at DeFiore and Company and to my amazing publicist Michelle Bega at Rogers & Cowan. Thank you to my incredible lawyer Jeff Bernstein, and to Cathey Lizzio, Pat Brady, and Matt Sherman for your patience and support. Thanks to Jamie B., Sonny P., and my dear friends Gocha & Shorena for the dance breaks, and to Hope Diamond, Brenda Grant, and Danielle Dusky, and my BFF, Kimmie. Thanks so much to Barbara Jacobson and everyone who proofread, including my mom, dad, Anne Clarke, Kay Carlson, Crystal, Kim, Brandy, Todd, Damon, Yvette, Kayo, Allen, Ryan, Benjie, Meeghan, and a very special thank-you to my amazing brother-in-law Mike Scafati and the incomparable, brilliant Jonathan Farley. Thank you to Nicole Cherie Jones, Brittany Pogue-Mohammed, Christina Tvardek, Ben Weiss, and Anne Lowney. Thank you to all who shared their stories, including my cousin Jazmin and my goddaughter Tori. Thank you to the educators who collected surveys, especially Dan Degrow, Kris Murphy, Joanne Hopper, Pete Spencer, Barb Treaster, and the staff at St. Clair RESA, and also to Shirley Stoll, JoAnn Aleman, and Lisa Miller. And thanks to teachers Matthew Hartman, Rachel Bosworth, Karen Boyk, David Fehringer, Ann Niedwiecki, and Kate Burnstein.

And finally, thank you to my loving sweetheart Michael. I'm at a loss for words to express my gratitude for your love, support . . . and great titles for my books!

Hot Topics

PART 5: EXPONENTS AND SQUARE ROOTS

PART 6: POLYNOMIALS

PART 7: QUADRATIC EQUATIONS

Hot . . . X?

In algebra, x can be anything: $\frac{1}{2}$, −47, a billion—the sky's the limit. And by the time you finish this book, you'll be a master at solving for x and a ton of other algebra topics. But can you imagine if when we solved for x, we started by stereotyping it, shutting out a whole range of possible solutions? Like, "Hmm, I just don't think x could be a very big number; I mean look, it's so little."

I know it sounds crazy, but this happens all the time in life. The x you're solving for is your future, and it, too, can be *anything*. But you'd be surprised how many people stereotype themselves, shutting out a whole range of possibilities for their future. I almost did. Although I got good grades, deep down I believed that no matter how many A's I might get, it was only a matter of time before I would fail and the truth would come out: I just wasn't a math person.

See, I had an image stuck in my head of who could be really good at math—nerdy guys who would grow up to look like Einstein—and I simply didn't *look* the part. I mean, I'd been an actress on television since the age of 12. I almost didn't take a math class in college at all, because I felt so intimidated. Luckily, I love a challenge so I went for it anyway. It floored me when I actually excelled! Today I have a degree in mathematics, and I've written three books about math. Just like in algebra, the most satisfying answers are often the ones we least expect.

I'm here to tell you that *giving up on ourselves* because of our own stereotypes and limited imaginations is a far more destructive force than any challenge or obstacle "out there." Just like in algebra, attaining x can be very challenging—but that's part of what makes reaching big goals so satisfying and valuable. And how do we strengthen the "not-giving-up" muscle? By doing things that require determination and stamina, like algebra.

Algebra can be difficult at times, but believe me, it's not out of your reach—at all. In fact, it will shape you into the kind of person who embraces challenges in all areas of life. Stick with me; you'll see how algebra is just an extension of what you already know. And just like that, the incredible woman you'll grow into will be a beautiful extension of the strong young woman you're crafting yourself into today.

By conquering *x*, you too can be anything. And *that's* hot.

FAQs: How to Use This Book

What Kinds of Math Will This Book Teach Me?

Algebra! The chapters of this book are filled with things like chocolate, puppies, birthday cake, pool parties, hair salons, jewelry, and lemonade. By the time you finish reading them, however, you'll be a whiz at tons of algebra topics, including polynomials, word problems, quadratic equations, graphing linear inequalities, and tons of new strategies for solving for x.

And just to make sure you're *never* confused, every single problem has an answer at the back of this book. There are also fully worked-out solutions on the "solutions" page of danicamckellar.com/hotx, so you can see *exactly* how to do them in case you get a different answer.

What Should I Already Know in Order to Understand This Book?

You'll want to have a pretty good handle on pre-algebra: factors, fractions, decimals, negative numbers, exponents, plotting points, etc. I'm the first to admit that the line between pre-algebra and algebra is pretty fuzzy, so this book picks up where my last book—*Kiss My Math: Showing Pre-Algebra Who's Boss*—ended. But, I mean, what are the chances you're a total expert on pre-algebra? Everyone forgets things!

To make sure that you never feel lost, throughout the book, I include just little bits of review and tons of footnotes that say stuff like, "To review such-and-such, see p. –– in *Kiss My Math* or even my first book, *Math*

Doesn't Suck," so you can flip to it quickly. If you don't own *KMM* and *MDS*, that's fine, too—there are other places to review those topics (by doing an online search for your topic). But this way, you're totally covered!

Do I Need to Read the Book from Beginning to End?

Nope! There are a few different ways to use this book:

- You can skip directly to the chapters that will help you with tonight's homework assignment or next week's test.

- You can skip to the math concepts that have always been problem areas to clear them up for good!

- Or you can, in fact, read this book from beginning to end and refer back to each chapter's "Takeaway Tips" for quick refreshers as you need them for assignments.

What's in This Book Besides Math?

In addition to the math I teach, look out for these fun extras, and more!

- Personality Quizzes: Are You Bold or Shy? Are You a Perfectionist? Find out now on pp. 72 and 348!

- Stories from teens just like you . . .

- A bonus feature on how to focus when distractions pop up!

- Real-life testimonials from gals who overcame their struggles in math and are now fabulously successful women! We've got a gourmet chef, a TV reporter/ actress, a daredevil airshow pilot, and more. And yes, they all use math in their jobs.

  Let's do it!

Chapter 1

Alge-blah

Up Close and Personal with Algebra

Algebra is nothing to be afraid of. Seriously, the word itself is the scariest part. So let's make a deal: Any time you start to feel queasy over the idea of algebra, I want you to say, "alge-blah-blah-blah. . . ." There are no limits to how many *blahs* you can use. Your turn. (Yes, out loud.)

But just to avoid any mishaps, for the next few pages I will replace the word "algebra" with the word "Happyland." Studies show that most people find this term more pleasant.* And please, do allow me to reintroduce you to our friend, *x*, just in case your first introduction was under less-than-desirable circumstances. A scary textbook, a grumpy teacher—these are all too often where we first meet little *x*, which is hardly fair to the tiny symbol that also means "kiss." Whenever you see *x*, or any of her friends like *y*, *z*, and so on, remember that she's just trying to be helpful, standing in for some number whose value we don't happen to know yet.

And because she's a placeholder, we can use any other letter we want, really. I've always found *n* to be a pretty friendly looking placeholder. You can even use other symbols, like □ 👝 ✿ ☺. Any time you're not in the mood for *x*, just replace it with any of these. I like to think of *x* as a box or a bag, because it looks like it could be "holding" the unknown value, y'know?

. . . .

* I didn't really conduct any studies. I'm playing the odds, okay?

And it reminds us why things like $2x + 3x = 5x$ are true:

$$2\square + 3\square = 5\square$$

Because after all, 2 boxes plus 3 boxes equals 5 boxes. But I predict that soon enough you'll warm up to x, and you'll be blowing kisses just like they do in Happyland.

Let's take a moment to review the basic players from pre-algebra.

What Are They Called?

Let's take a look at this little math expression and name its parts:*

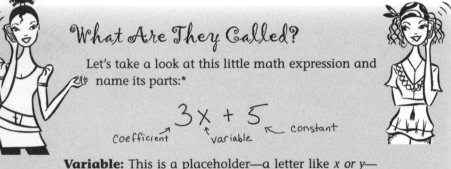

$$3x + 5$$

Coefficient   variable   constant

Variable: This is a placeholder—a letter like x or y—that stands in for a number we don't know yet. It could have a variety of different values; hence the name *variable*—for, um, variety. Sometimes we solve an equation to find out the variable's value, but in expressions like this one, the variable's value remains a mystery . . . forever.

Constant: This is something that will make you feel all warm and mushy inside, reminding you of kindergarten when we just had numbers by themselves. In the expression $3x + 5$, the 5 is constant, unchanging, dependable. Ah, how nice. By the way, in this expression, the 3 is *not* considered a constant, because it's stuck to the variable. The 3 will tell us how many of the mysterious x we have.

Coefficient:[†] Even though we don't know the value of x, the coefficient is the number that will tell us exactly *how many* of this totally unknown thing we have. So in the expression $3x + 5$, the 3 is the coefficient. We might not know what x's value is, but we know we have 3 of them! By the way, $3x$ is a more efficient way of writing $3 \cdot x$ and a much more efficient way of writing $x + x + x$, don't you agree? Coefficient: so efficient!

. . . .

* And a more in-depth version of this can be found on pp. 88–89 of *Kiss My Math*.
† Some textbooks call this a *numerical coefficient*.

Terms: These are numbers and/or variables that are *stuck together* with multiplication and division (as fractions), and separated from each other with addition and subtraction. So, $3x + 5$ has two *terms* in it, and $\frac{x}{2} + 8 - 2xy - 3$ has four *terms* in it. If you're close enough to spread a *germ*, you're part of the same *term*!

Phew! Now that we have the players in place, we're going to broaden your perspective of what you already know from pre-algebra. And I'm going to let you in on a secret: The recurring pre-algebra themes below are the backbone of math, in algebra (um, Happyland) and beyond.

1. Writing the same value in different ways
2. Translating between English and the language of math

Get good at these, and you'll totally kick butt on tests and homework.

Same Value, Different Look!

I can just hear them now: "You look different! Have you been working out? New haircut, perhaps?" Ever notice how when someone's in a really good mood, she just glows? She *looks* different, even though nothing has really changed except her attitude. The same thing happens in Happyland: We can give math expressions a whole new *look* without changing their values at all. For example:

$$10 - 3 \text{ has the same value as } 7.$$

$$\frac{1}{3} \text{ has the same value as } \frac{10}{30}.$$

$$5x \text{ has the same value as } \frac{5x}{1}.$$

Some people call this sort of thing *renaming*, which has always seemed like a pretty funny word to me. It makes me think of somebody who stops calling her dog Sparky and starts calling him Fido instead.

I've collected a whole bunch of familiar renaming/rewriting tricks to make sure you're in tip-top shape; we'll be using all of 'em throughout the book. And I'll tell you where to go in case you want to brush up a little. Believe me, reviewing pre-algebra is never a waste of time!

Renaming Review: Same Value,

Category	One way to express this value	equals	...another way to express the same value!
Equivalent fractions	$\dfrac{2}{6}$	=	$\dfrac{1}{3}$
Whole numbers as fractions	7	=	$\dfrac{7}{1}$
Decimals, fractions, percents	0.6	=	$\dfrac{3}{5}$ or 60%
Subtraction vs. "adding a negative"	$C + (-3)$	=	$C - 3$
All variables have coefficients	x	=	$1x$
Multiplying times a fraction	$\dfrac{1}{2}w$	=	$\dfrac{w}{2}$
Cruel pet owner	Sparky	=	Fido
All variables have exponents	Y	=	Y^1

Different Look!

More details

We can rename or rewrite any fraction by multiplying or dividing the top and bottom of the fraction by the same thing. So, $\frac{2}{6} = \frac{2 \div 2}{6 \div 2} = \frac{1}{3}$. And $\frac{2}{6} = \frac{2 \cdot 5}{6 \cdot 5} = \frac{10}{30}$.* They've all got the same <u>value</u>.

This trick can be very handy for finding reciprocals,† among other things. We can put *anything* on top of a 1, and its value is unchanged, so $0.02 = \frac{0.02}{1}$ and $n = \frac{n}{1}$.

Moving between these three forms for the same value comes up all the time, especially in word problems.‡

We can rewrite any subtraction as "adding a negative," and vice versa. See the next page for more!

"Invisible" coefficients equal 1 or −1. So the expression x − y can be rewritten as 1x − 1y.

Multiplying a number or a variable times a fraction is the same as multiplying times just the numerator (top) and keeping the denominator (bottom). See the next page for more!

If you don't see an exponent on a variable, that means the exponent is 1. Easy, right? We'll do more with exponents in Chapters 17 and 18.

---

* To brush up on equivalent fractions, see Chapter 6 of *Math Doesn't Suck*.
† See p. 43 for reciprocals.
‡ You can review this in Chapters 11–14 of *Math Doesn't Suck*.

As you can see, we have many choices about how to write our expressions, depending on our mood (or whatever makes the homework problem easier). Let's go over a couple of these renaming tricks in more detail.

"I used to be afraid of math. Now I realize that math should be afraid of me." **Jessica, 13**

Streamlining Negatives: $b + (-3) = b - 3$

When first learning about negative numbers, it can be really helpful to rewrite subtraction as "adding a negative," so rewriting $5 - 7$ as $5 + (-7)$ makes it easier to see that the answer is **−2**. This is a good trick for working with variables too, especially when we're dealing with *like terms*:

$$4g + 5h - 3g - 7h$$

It's much easier to combine terms if we rewrite the subtraction as "adding negatives":*

$$4g + 5h + (-3)g + (-7)h$$

And combining the coefficients, we get **$1g + (-2)h$**.

However, for final answers, it looks neater to streamline the negatives back into subtraction and also to take off the 1 coefficient: **$g - 2h$**. See what I mean? Much tidier.

Multiplying Times a Fraction: $\frac{1}{2}w = \frac{w}{2}$

As it turns out, multiplying anything times a fraction is the same as multiplying times just the numerator (top) and keeping the same denominator (bottom). Ever thought about it like that? For example, in order to multiply 5 times $\frac{2}{3}$, we just multiply 5 times the 2 and keep the

. . . .

* For combining negative numbers (and breath mints!), see Chapter 1 in *Kiss My Math*, and for combining *like terms* with variables, see Chapter 9 in *Kiss My Math*.

denominator: $5 \cdot \frac{2}{3} = \frac{5 \cdot 2}{3} = \frac{10}{3}$. It's easy to see why this is true if we write 5 as a fraction, because then $5 \cdot \frac{2}{3}$ becomes $\frac{5}{1} \cdot \frac{2}{3}$, and normal fraction multiplication* gives us the answer: $\frac{10}{3}$.

The same is true for variables. So, $\frac{1}{2} \cdot w$ (or $\frac{1}{2} w$) can be rewritten as $\frac{w}{2}$. Just like with numbers, it's easier to see why this is true if we first rewrite w as a fraction: $\frac{1}{2} w = \frac{1}{2} \cdot \frac{w}{1} = \frac{w}{2}$. Changing the position of the variable from the outside to the top of the fraction, and vice versa, can be very helpful when we get into factoring and other things, too. Here are some more examples of this:

$$3 \cdot \frac{a}{2} = \frac{3a}{2} \qquad (-g) \frac{-1}{5} = \frac{g}{5} \qquad \frac{10x}{7} y = \frac{10xy}{7}$$

It's just good to be flexible!

When you were little, did you push your vegetables to one side of your plate so it looked like you had eaten some of them? (I used to do this with my peas.) Well, your mom knew you hadn't eaten any of them, but there you were, being a little mathematician and not even knowing it; you were expressing the *same value* in a different way. Too bad it couldn't get me out of eating my peas. Yeah . . . she caught on.

More pre-algebra tricks like the commutative property, the associative property, and the distributive property allow us to rewrite things, too. At the moment, I don't think you need me to remind you that $a + b$ is the same as $b + a$, but check out pp. 399–400 in the Appendix to brush up on these tricks if you want—the distributive property is especially helpful.

Again, one of the secrets to succeeding in math is knowing how to rewrite stuff to our advantage, so let's warm up!

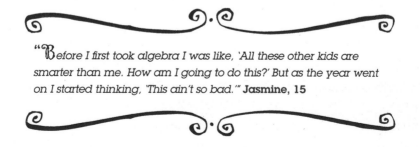

"Before I first took algebra I was like, 'All these other kids are smarter than me. How am I going to do this?' But as the year went on I started thinking, 'This ain't so bad.'" **Jasmine, 15**

. . . .

* To brush up on fraction multiplication, see Chapter 5 in *Math Doesn't Suck*.

Doing the Math

Match up the equivalent expressions. I'll do the first one for you.

1. $\frac{2}{3}b$

<u>Working out the solution</u>: We know we can rewrite this with b on top without changing the value: $\frac{2b}{3}$. Hmm, it doesn't match A, because b appears on the bottom of the fraction, which changes the value. But E can be rewritten like this: $\frac{4b^1}{6} = \frac{4b \div 2}{6 \div 2} = \frac{2b}{3}$. We've found a match!

Answer: E

2. $\frac{4}{6b^1} - 1 + \frac{a}{a}$

3. $b - a$

4. $\left[\frac{a}{1} + (-1b) \right]$

5. $-[1 - a - b] + 1$ *(Hint: Distribute that negative sign!)*

(Answers on p. 402)

A. $\dfrac{2}{3b}$

B. $1(a^1 + b)$

C. $a - b$

D. $(-1)a + 1b^1$

E. $\dfrac{4b^1}{6}$

Translating Between English and Math: Parlez-vous Math?

Make no mistake about it: Math is a foreign language! And translating between English and math is a great skill to sharpen.

X, a Know-It-All's Best Friend

Have you ever known someone who never admits when she doesn't know something, so she just makes something up? "Why yes, the capital of Peru is Soybean."*

In Happyland, we never admit it when we don't know a number. But instead of inserting a *wrong* number, we insert a placeholder, like *x*. So, suppose someone says, "Eight is one-half of what?" Without admitting we don't know, we just say, "Why yes, 8 is $\frac{1}{2}$ of *x*." See? We sound like a know-it-all.

Let's see how this looks, translated:

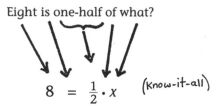

Eight is one-half of what?

$$8 \quad = \quad \frac{1}{2} \cdot x \qquad \text{(know-it-all)}$$

Now that we've translated English into math, it's ready to be solved. We just multiply both sides by 2 and get **x = 16**.

QUICK NOTE   A pre–algebra reminder: When translating from English to math, "*is*" becomes =, and when "*of*" is immediately surrounded by two numbers or variables (like in the example above), it becomes *multiplication*.†

Let's face it; nobody likes a know-it-all in life, but a secret know-it-all attitude is actually very empowering in math, especially when solving word problems. Any time there's a value we don't know, we just boldly pick a variable or a placeholder to stand in for that value (also called *labeling*), and proceed with confidence to write a true math sentence—that we can then solve.

. . . .

* The capital of Peru is Lima. Right idea, wrong bean.
† For many more translation tips, check out Chapter 11 in *Kiss My Math*.

So you see, *x* is a tool*—a friendly helper to make problems easier to solve, allowing us to translate from English into math whenever we don't know something. It has other uses, too: In Chapter 6, I'll show you a fun way to use *x* to make your holiday shopping totally stress-free!

"Omigod, I'm Soooo Bad at Reading."

*Y*eah, you probably don't hear that too often, especially not from adults. After all, reading comes up in everyday life, and not being able to read would really limit a person's ability to take care of herself, right? So then, why are so many people ready to "admit" they're terrible at math?

After all, math is also essential for taking care of ourselves, especially when it comes to money. Being able to calculate interest rates on credit cards, learning to budget, understanding your taxes, and balancing a checkbook are important for managing money and staying out of debt as an adult, which is more important than you might realize. It's not just fabulous women running their own businesses who need math—it makes a real difference for all of us.

Just keep up all your hard work in math: Stay sharp and savvy, and when you become an adult, you can be in much better control of your finances—and your life!

Consecutive Integer Problems

These problems happen to be very popular with teachers and on standardized tests. They're also good for warming up our know-it-all labeling muscles.

. . . .

* As in "used as a means of accomplishing a task," not as in "total loser."

What's It Called?

Consecutive means "following each other in order." For example, instead of saying "we had three wins in a row," we could say, "Our team has had three *consecutive* wins!"

The **integers** are the whole numbers and their negative counterparts, not including any fractions or decimals {. . .–3, –2, –1, 0, 1, 2, 3 . . .}. (See pp. 398–99 for more number sets.)

So **consecutive integers** are just integers, in order!

For example, the first five consecutive positive integers are 1, 2, 3, 4, 5. The first five consecutive *odd* positive integers are 1, 3, 5, 7, 9, and the first five consecutive *even* positive integers are 2, 4, 6, 8, 10. Make sense?

Say we want to find two consecutive integers that add up to equal 35. We could start by labeling the first number n and the next number $(n + 1)$. Do you see why? Even though we don't know n's value, we do know that $(n + 1)$ will definitely be the next number after it, right? And since we know the sum of n and $(n + 1)$ equals 35, that means our math sentence would read:

$$n + (n + 1) = 35$$

Some number · · · the next number

From there, we could solve it! $2n + 1 = 35$ (subtracting 1 from both sides) → $2n = 34$ (dividing both sides by 2) → $n = 17$. And that means $n + 1 = 18$, so the answer is **17 and 18**. Here are some helpful labeling tips:

> **Two unknown consecutive odd integers** can be labeled n and $(n + 2)$, because they will be two away from each other, just like 1 & 3 or 7 & 9. Makes sense, right?
>
> **Two unknown consecutive even integers** can *also* be labeled n and $(n + 2)$, because they will also be two away from each other, just like 2 & 4 or 8 & 10.
>
> **Three unknown consecutive integers** can be labeled n, $(n + 1)$, $(n + 2)$. Each expression is 1 more than the previous one!

QUICK NOTE   It's usually easiest to label the smallest unknown integer *n* and go from there. And remember, integers can be negative. I'll do an example below in problem 1. And always look out for the words *odd* and *even*.

If you're feeling unsure about how you've labeled your unknown integers, just plug in any appropriate number for *n* and see if it works. So if you've labeled two unknown consecutive odd integers *n* and *n* + 2, just plug in an odd integer for *n*, like for example 3. For *n* + 2, you'd get **3** + 2 = **5**. And yep, 3 & 5 are two consecutive odd integers!

Doing the Math

a. Translate these consecutive integer problems into math language, using *n* to label the smallest integer you're looking for.

b. As a bonus, solve the problem!* I'll do the first one for you.

1. The sum of three consecutive even integers is –30. What are they?

<u>Working out the solution</u>: If we label the smallest one *n*, then the next even integer would be *n* + 2, and the one after that would be *n* + 4. So far, so good? And we're told this must all add up to equal –30, so that would translate into math like this: **n + (n + 2) + (n + 4) = –30.** For part b, let's drop the parentheses and combine terms to get
3n + 6 = –30   →   3n = –36   →   n = **–12.** This

....

* To brush up on negative numbers and solving for *x* (from pre-algebra), check out Chapters 1, 7, 9, and 11 in *Kiss My Math*.

means $n + 2 = -10$ and $n + 4 = -8$. And yep,
$-12 + (-10) + (-8) = -30$. We got it right!

Answer: a. $n + (n + 2) + (n + 4) = -30$ b. -12, -10, -8

2. Find two consecutive integers whose sum is –11.

3. The sum of three consecutive integers is 42. What are they?

4. Find three consecutive odd integers whose sum is –3.

(Answers on p. 402)

QUICK NOTE   For tons of practice translating from English to math in other types of math problems, check out Chapter 11 in *Kiss My Math*. It'll help get you ready for the word problems in Chapters 13–16 of this book.

Takeaway Tips

 Don't let the idea of algebra intimidate you! Just think "Alge-blah blah blah."

 Variables are just placeholders for some number we don't know yet.

 There are many ways to write the same value in math! Getting good at changing the look of an expression without changing its value is one serious key to success in math.

Math is a language. And translating between English and math is a great skill to sharpen for algebra, especially for word problems.

Review Like the Pros!

Did you know that the best professional dancers spend time just balancing on one leg, because they know it will improve their pirouettes? Soccer players do drills running down the field—sometimes without a ball—because they know it will improve their game. Reviewing topics in pre-algebra, like fractions or solving for x, will totally improve your performance in complicated algebra problems. You'll be amazed at the new stuff you'll finally understand. The pros know it: Going back to the basics is never a waste of time, and it'll set you apart from the crowd, missy!

Chapter 2

Happy Birthday to Me

GCFs and LCMs with Variables

When you blow out your birthday candles, what do you wish for? When I was in high school, I usually wished for some guy that I was crushing on to like me back. Hmm. Come to think of it, I don't think that ever worked.

Have you ever wished for math to get easier? Well, that's about to come true, even if you never wished for it. See, for the next few chapters, we're going to do things that you already know how to do. The only difference is that we're going to add variables into the mix. And since variables are just numbers whose values we don't know yet, it'll all work pretty much the same way.* Oh, and there will be birthday cake later in the chapter, just to round out the whole birthday experience.

First, let's brush up on a few familiar definitions.

What Are They Called?

Product: The <u>result</u> of things being multiplied together: 6 is the *product* of 2 & 3, and 4*x* is the *product* of 4 & *x*.

Factor: Anything that is being multiplied by something else to create a product. So, 2 & 3 are both *factors* of 6, and 4 & *x* are both *factors* of 4*x*.

. . . .

* And if you need to brush up on how to do this stuff with numbers, I'll always tell you where to find friendly help in *Math Doesn't Suck*.

Product of factors: This is just a group of stuff stuck together by multiplication.* For example, $5xy$ and $\frac{z}{2}$ are both *products of factors*.

Prime factor: This is a *factor* that also happens to be a prime number *or a single variable with no exponent*. For example, in the expression $21d$, the prime factors are 3, 7, and d.[†] Prime factors have a lot in common with monkeys. (Stay tuned.)

QUICK NOTE   Exponents: In Chapters 17 and 18, we'll learn all about exponents. Until then, just remember from pre-algebra that b is the same thing as b^1, and also that $b \cdot b = b^2$, and $b \cdot b^2 = b^3$.

Finding the Greatest Common Factor (GCF)

Do you remember finding the *greatest common factor* of two numbers?

The GCF is the biggest factor that divides evenly into both numbers. So, somebody might say, "Find the GCF of 24 and 30." One way to figure this out would be to draw factor trees, circle their *prime* factors (the prime factors swing off the lowest branches

. . . .
* Or division in fraction form, because after all, division is the same as multiplication by the reciprocal. (See p. 43 for more on this.) It all depends on how you look at it!
[†] In case you're wondering, it doesn't matter if d turns out to be a prime number. All that matters is that for as long as we don't know the value of d, we consider it prime because we simply don't know how to split it up into smaller factors. On the other hand, d^2 would not be a prime factor because we can write it as $d \cdot d$.

of factor trees, much like monkeys), underline the ones they have in common, and multiply them together:*

Remember, each time you draw branches from a number, ask yourself the question: "What two factors will give me this number when multiplied together?" We usually have a few choices for how to start, but if we do it right, the *prime* factors at the end will always be the same. Monkeys are very dependable like that.

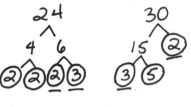

Common prime factors: **2, 3**

GCF of 24 and 30 is $2 \cdot 3 = $ **6**

Well, the same method works with variables. If we want to find the GCF of **2*ab*** and **6*b*²**, we could draw factor trees and circle their monkeys (um, prime factors) and underline the ones they have in common. The product of those common prime factors will be the GCF. And each time you draw branches extending outward, ask yourself the question, "What two factors will give me this expression when multiplied together?"

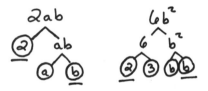

Common prime factors: **2, *b***

GCF of 2*ab* and 6*b*² is $2 \cdot b = $ **2*b***

We've multiplied all the prime factors that they had in common, so we've found the biggest thing that is a factor of both 2*ab* and 6*b*²: their GCF.

· · · · · · · · · ·

* This should look pretty familiar. If not, you can check out Chapters 1 and 2 in *Math Doesn't Suck* for a review of factor trees and monkeys, I mean, um, *prime-ates.* Get it?

What's It Called?

The **greatest common factor** (GCF) of two expressions is the biggest product of prime factors that they share in common. In other words, it's the biggest thing that divides into both of them evenly.

For example:

The GCF of 18 and 24 is **6**.   The GCF of $2ab$ and $6b$ is **$2b$**.

The GCF of $2h$ and $3g$ is **1**.   The GCF of $6y^2$ and $9xy^2$ is **$3y^2$**.

Notice that if two expressions have no factors in common (except 1), then the GCF of the two expressions is 1.*

As cute as the monkeys are, it can be annoying to write out all those variables like we did on p. 17, especially when exponents get involved. And as if someone answered our birthday wish, here comes a shortcut!

Shortcut Alert

Handle coefficients and variables separately, and compare powers!

Instead of writing out all those variables and underlining the common ones, try this: First, just find the GCF of the coefficients (the numbers), and then compare the *exponents* on the variables the terms share.

For example, to find the GCF of $30xy^3$ and $24x^2y^2$, we'd pick our favorite way to find the GCF of their coefficients (30 and 24), which is **6**,† and then for each variable they share, *compare exponents*. The variable with the *lowest* exponent wins each time:

Compare x and x^2; **x** is lower, so it wins!

Compare y^3 and y^2; **y^2** is lower, so it wins!

---

* Just like with numbers, these two expressions would then be called *relatively prime*.
† You could do factor trees to find the GCF of 30 and 24, or see Chapter 2 in *Math Doesn't Suck* for more ways to find GCFs of numbers.

And the GCF is just the product of the coefficients' GCF and these variable "winners."

$$6 \cdot x \cdot y^2 = \mathbf{6xy^2}$$

Ta-da! By the way, if you took the time to draw their factor trees, you'd get this same answer. But really, who has time for all those monkeys?

Comparing exponents on the variables makes sense when you think about it: The *lowest* power of each variable they share is *guaranteed* to be a factor of both terms.

Step By Step

Finding the GCF of two (or more) expressions:

Step 1. Look at the coefficients—just the numbers—and find their GCF. You can use a factor tree or any other method you like.

Step 2. For each variable that the terms have in common, *compare exponents*. The lowest exponent wins! Write down the list of winners.

Step 3. Multiply the coefficients' GCF times the variable "winners." That's the entire GCF. Done!

And... Action! Step By Step In Action

Find the GCF of $51c^4d$ and $9c^2d^6e$.

Step 1. First, we'll find the GCF of 51 and 9. A good 'ol divisibility trick tells us that since $5 + 1 = \mathbf{6}$, which is divisible by 3, that means 51 is divisible by 3, too.* In fact, $51 = 3 \cdot 17$ (and 17 is prime). So the GCF of 51 and 9 is **3**.

. . . .

* This "adding up the digits" trick only works for 3 and 9. Check out more divisibility tricks on p. 9 of *Math Doesn't Suck* or online at danicamckellar.com/MDS/extras.

Step 2. Next, let's look at the variables that $51c^4d$ and $9c^2d^6e$ share in common: c and d.

For c, we'll compare c^4 and c^2; **c^2 wins!**

For d, we'll compare d and d^6; **d wins!**

Step 3. So our GCF is $\mathbf{3 \cdot c^2 \cdot d = 3c^2d}$.

Answer: $3c^2d$

QUICK (REMINDER) NOTE   You can always write in the invisible 1 exponents on variables, so: $a \rightarrow a^1$.

"*I* don't think most teenage girls realize that 'smart' is what lands you the job, the money, and the life that a lot of us dream of for the future." **Jade, 15**

Doing the Math

Find the GCF of the following pairs of expressions. I'll do the first one for you.

1. $5hm^2k^3$, $9hkm^3$

<u>Working out the solution</u>: To help keep the variables straight, let's put each term in alphabetical order: **$5hk^3m^2$, $9hkm^3$**. Okay, since 5 and 9 are relatively prime, their GCF is **1**. First, we compare h to h; so **h** "wins" and will be part of the GCF. Next, we'll compare k^3 and k; **k wins!** Finally, we'll compare m^2 and m^3; **m^2 wins!** Now, we just multiply them together: **$1 \cdot h \cdot k \cdot m^2 = hkm^2$**. Done!

Answer: hkm^2

2. $4a^2b$, ab^2

4. $5g^2h^4$, 15

3. $14x^3y^2$, $28xy^2z$

5. $6a^2b$, $4ba^2$

(Answers on p. 402)

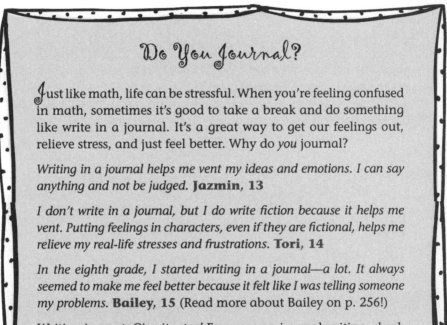

Do You Journal?

*J*ust like math, life can be stressful. When you're feeling confused in math, sometimes it's good to take a break and do something like write in a journal. It's a great way to get our feelings out, relieve stress, and just feel better. Why do *you* journal?

Writing in a journal helps me vent my ideas and emotions. I can say anything and not be judged. **Jazmin, 13**

I don't write in a journal, but I do write fiction because it helps me vent. Putting feelings in characters, even if they are fictional, helps me relieve my real-life stresses and frustrations. **Tori, 14**

In the eighth grade, I started writing in a journal—a lot. It always seemed to make me feel better because it felt like I was telling someone my problems. **Bailey, 15** (Read more about Bailey on p. 256!)

Writing is great. Give it a try! For more on journal writing, check out p. 122 and p. 282.

Multiples—with Variables!

As you may recall, *multiples* of numbers are easy to find. Here are some multiples of 3:

$$3 \times 1 = \textbf{3}, \quad 3 \times 2 = \textbf{6}, \quad 3 \times 3 = \textbf{9}, \quad 3 \times 4 = \textbf{12} \ldots$$

Finding multiples of a variable like *n* works the same way:

$$n \times 1 = \textbf{\textit{n}}, \quad n \times 2 = \textbf{2\textit{n}}, \quad n \times 3 = \textbf{3\textit{n}}, \quad n \times 4 = \textbf{4\textit{n}} \ldots$$

Believe it or not, you'll eventually find that multiples of variables are actually *easier* to deal with than multiples of numbers for the same reason that 18×7 is easier to work with than 126. When some of the factors stay "spelled out" for us, it's so much nicer than dealing with a big, messy number.

Least Common Multiple—with Variables!

I'm sure you remember the delights of adding and subtracting fractions like this: $\frac{7}{8} - \frac{5}{12}$. In order to subtract these, the first step is to convert them into equivalent fractions with common denominators, right? Because we all like small numbers, we use the *smallest* denominator we can—in other words, the **least common multiple*** (LCM) of 8 and 12, which is **24**.

In Chapter 5, we'll use LCMs in order to add and subtract fractions with variables in their denominators. So let's get good at finding these LCMs!

Ring Ring ~ What's It Called?

The **least common multiple†** (LCM) of two expressions is the smallest expression that has both original expressions as factors. For example:

The LCM of 8 and 12 is **24**

(. . . because 24 is the lowest thing that has both 8 and 12 as factors).

The LCM of xy and x^2 is x^2y

(. . . because x^2y is the lowest thing that has both xy and x^2 as factors).

The LCM of $8xy$ and $12x^2$ is $24x^2y$

(. . . because $24x^2y$ is the lowest thing that has both $8xy$ and $12x^2$ as factors).

And how do we find LCMs? I'll give you a hint: "Happy birthday to me . . ."

. . . .

* To review finding the LCMs of numbers, see Chapter 3 in *Math Doesn't Suck*.
† This is sometimes called the **lowest common multiple**.

The Return of the Birthday Cake

In *Math Doesn't Suck*, I showed you this method, which ends up looking a little like a birthday cake . . . sort of. Now we'll do it with variables! Let's find the LCM of $3xy$ and $12y$. First, we draw a shelf or "layer" under them, and we look for a common factor—any common factor.

How about 3? So we write the 3 on the left, divide 3 into each term, and write the leftovers underneath: xy and $4y$. Since the leftovers still have a common factor, y, we draw another cake layer. Now we factor out the y, write it on the left, and our new leftovers are x and 4. Because x and 4 have no common factors, we're done drawing layers!

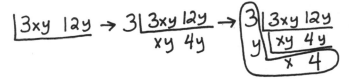

The LCM is found by multiplying everything along the big "L": $3 \cdot y \cdot x \cdot 4 = \mathbf{12xy}$. And as a bonus, the GCF of the two original terms can be found by multiplying just the left side: in this case, $3 \cdot y = \mathbf{3y}$. Well, happy birthday to all of us.

Step By Step

Using the Birthday Cake method for GCFs and LCMs of two terms:

Step 1. Write the two terms and draw a little shelf or "cake layer" under it.

Step 2. Now pick a factor—any factor—that both of them share. It can be a number, a variable, or a product of both. Write this factor on the *left* side of the layer. Divide the factor into each term, and then put each answer directly underneath. <u>Make sure that if you were to multiply this new layer times the factor you just used, you'd get the layer directly above it.</u> That's how you know you did it right.

Step 3. If the new layer has common factors, draw another "shelf" and repeat Step 2. Continue until you end up with a layer that has no common factors to all of its terms.

Step 4. Now that the bottom layer has no more factors in common, take the factors along the left side, and multiply them together—that's your GCF.

Step 5. To find the LCM of the terms, take all the stuff along the side and bottom—the big L—and multiply them together. That's your LCM!

And... Action! *Step By Step In Action*

Find the GCF and the LCM of $3ab^2$ and $6b$, using the Birthday Cake method.

Steps 1 and 2. First, we'll draw a cake layer underneath the two terms. Sure, we could factor out a 3 first, but just for fun, let's factor out a **b** first; that leaves us with **3ab** and **6**, right? So we'll put those on the next cake layer.

Step 3. Our new layer has a factor of **3** in common, so let's factor that out, and we'll end up with just **ab** and **2**. (Yep, if we multiplied this newest layer times 3, we'd get the layer directly above it.) Since ab and 2 don't have any common factors, it's the final layer.

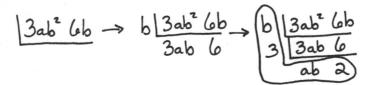

Step 4. To find the GCF, we just multiply along the left: $b \cdot 3 = 3b$. And this makes sense; it's all the stuff that the original factors had in common: the *greatest common factor*.

Step 5. To find the LCM, we multiply along the big L:
$b \cdot 3 \cdot ab \cdot 2 = 6ab^2$.

Answer: The GCF is 3b, and the LCM is $6ab^2$.*

. . . .

* Note: If we wanted to add or subtract fractions with denominators of $3ab^2$ and $6b$, we'd use **$6ab^2$** as a common denominator. (We'll do this on p. 64.)

Watch Out!

Although the Birthday Cake method can find the GCF for as many terms as you want (as we'll do in Chapter 3), the Birthday Cake method finds the <u>LCM</u> for only *two* terms at a time. If you lined up three terms and tried to find their LCM with cake, you could get the wrong answer! So, if you want to find the LCM of three terms, first find the LCM of *two* of the terms, take *that answer*, and find the LCM of *that answer* and your third, unused term. The final LCM will actually be the LCM of all three terms. See below for an example.

Take Two: Another Example

Find the LCM of 3xy, 12y, and $5x^2$.

Since we can only use the Cake method on *two terms at a time for LCMs*, first let's do the Cake method with just $3xy$ and $12y$. Well, we did that on p. 23, and found that the LCM of $3xy$ and $12y$ is **12xy**.

Next, we do the Cake method on this answer, **12xy**, and the third, unused term, $5x^2$. That final answer will be the LCM of all three terms.

Steps 1 and 2. We'll draw our layer, and hmm, the only factor $12xy$ and $5x^2$ have in common is x, so we'll write that on the side, and below we'll write the leftovers: $12y$ and $5x$.

Notice that if we multiplied that outside x times each of the new layer terms, $12y$ and $5x$, we'd end up with the layer directly above it, $12xy$ and $5x^2$, just like we should.

Steps 3–5. Our new layer has no factors in common, so that's the last layer. The LCM of $12xy$ and $5x^2$ is the product of the big L: $x \cdot 12y \cdot 5x = \mathbf{60x^2y}$. And that's the LCM of all three terms!

By the way, it didn't matter which two terms we started with; we could have found the LCM of $3xy$ and $5x^2$ first, which is $15x^2y$. Then we could have found the LCM of $15x^2y$ and the unused term, $12y$, which is indeed **$60x^2y$.**

Answer: The LCM of $3xy$, $12y$, and $5x^2$ **is $60x^2y$.**

Doing the Math

Find the GCF and the LCM of the following expressions. I'll do the first one for you.

1. $15ab$ and $8c$

<u>Working out the solution</u>: So we draw our layer, but this is odd—there are no factors in common. Let's try the cake anyway. Since the only factor they share is 1, we'll write that on the side, which means we just copy the top layer down below.

$$1 \,\lfloor\, \underline{15ab \quad 8c}$$
$$15ab \quad 8c$$

The GCF is the left side: 1. To find the LCM, we just multiply the L, which is: $1 \cdot 15ab \cdot 8c = $ **$120abc$**

Answer: The GCF is 1, and the LCM is $120abc$.

2. $7jk$ and jk^2

3. $33m^2n$ and $22mn^2$

4. $4ab$, $6a^2b$, and $9b$ (Hint: Find the GCF of all three at once. Then as a <u>separate problem</u>, do the LCM of the terms two at a time, like I did on p. 25 above.)

(Answers on p. 402)

Takeaway Tips

 The greatest common factor (GCF) of two or more terms is the biggest product of the prime factors that they all share. Using the Cake method, the GCF can be found by multiplying the terms on the left side of the cake.

 The least common multiple (LCM) of two or more terms is the smallest expression that has those terms as factors. Using the Cake method, the LCM for two terms can be found by multiplying everything along the big L, so it's easy to remember!

 To use the Cake method to find the LCM of more than two terms, you must do them two at a time.

Danica's Diary
ROMANCE AND *TWILIGHT*

As a teenager, I remember having the strongest feelings surging inside me and wanting so desperately to run into the arms of a knight in shining armor. It didn't even matter who the guy was, really—the emotions were already in place. To directly quote from my diary at age 14, "Feelings are rushing through my body like a waterfall and distracting me from my homework and everything else! Ah! I just want to *kiss* someone!"

I didn't act on those feelings, but sometimes it felt like they consumed me. Where do these feelings come from? They're biological hormones, just like how testosterone makes most teenage guys want to have sex and play violent video games. We girls have our own strong hormones, too, but they make us crave other things: romance, reckless abandon, and being "whisked away" somehow. So, watching a movie like *Twilight* totally indulges these feelings, just like the violent video games do for guys. But let's not get carried away. I mean, these movies are just fiction, and do make the girls look pretty needy and pathetic, you have to admit. And here's a little reality check: If a guy is aloof and cold, he's not actually desperately in love with you. He's, um, probably not interested. And, oh yeah, if a guy says he might hurt you or that he's "dangerous"—seriously, run! Let some other girl be the victim of his abuse. You're way too smart to let your hormones make important decisions for you, little missy.

TESTIMONIAL

Catherine Tran Fellows
(Brooklyn, NY)
<u>Before</u>: Struggled with math
<u>Today</u>: Director of a modeling agency!

Everyone assumes Asians are good at math, but I struggled with it starting in grammar school. Because of the stereotype, I thought I "should" be good at math, and I often felt embarrassed at my lack of natural talent. My parents tutored me at home, doing plenty of math outside of my school assignments. Even though it was difficult for me and I felt a lot of pressure, I worked hard, and my math skills did start to improve in high school. Now I'm grateful that math didn't come easily, because I really believe that struggling in math actually forced me to develop other strengths, too, such as leadership skills and social skills. I wasn't going to just be in the corner somewhere, doing perfect math by myself. I needed to be social, interactive, join study groups, and ask others for help. And as it turns out, the confidence that I built led me to a really fun, social career, where I use math every day!

As the director of a modeling agency, I represent photographers and models. I love traveling to foreign countries to scout for models. Finding new faces and developing them to become successful models is so fun and rewarding. And the red carpet events are fun, too!

Behind all the glamour, a modeling agency is a business like anything else. For example, I calculate hourly pay for my part-time employees manually. And if an employee works a fraction of an hour, I have to use fractions to figure out his or her wages. I also work with percents constantly: I deduct 20% commission from the models who work for me. I pay my salary employees 5% commission on their sales. And negotiating contracts can be a lot like a big word problem in algebra, coming up with numbers and logistics that make both sides happy.

I'm grateful for math in the volunteer work that I do, too. I'm the treasurer for the McCarren Dog Park and on the fundraising committee for the New York Junior League, where we organize social events to raise money for charity. We're an all-women organization full of leaders in volunteerism, and 85% of us have full-time jobs. It's a great feeling to be surrounded by such generous, talented, strong women and to consider myself one of them.

And the fundraising we do? It's all about the numbers.

Chapter 3

Listen to Your Gut

Factoring with Variables

$\mathcal{H}$ave you ever found yourself at a party, and it's just not your vibe? Maybe the parents aren't home, strange older kids are there, or worse, something dangerous is going on like drugs? For whatever the reason, big or small, you've realized it's time to pull out of the party. This instinct could save your butt . . . and oddly enough, it's going to help you factor with variables, too.

In *Kiss My Math*, we talked about how when you first arrive at a party, you probably want to say hello to everyone—and that's how we can think about the **distributive property**. The parentheses are like the walls of the house, and you're on the outside, ringing the doorbell. When someone opens the door, it's time to say hello!

You, saying "hi"!

$$a(b + c) = ab + ac$$

Ring Ring What's It Called?

The Distributive Property (of multiplication over addition and subtraction)

The *distributive property* is a rule that allows us to rewrite specific kinds of expressions (like the ones on the next page) so the parentheses go away but so the value of the expression remains unchanged. Imagine arriving at a party and saying hi to each friend!

For any real numbers a, b, and c, here's how it works.

With addition: $a(b + c) = ab + ac$

With subtraction: $a(b - c) = ab - ac$

This also works when the individual terms are a little more complicated, like this:

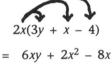

$$2x(3y + x - 4)$$

$$= 6xy + 2x^2 - 8x$$

The $2x$ says hi together, because it's considered to be a single *term*.*

Okay, now imagine you got to the party, you've just said hi to everyone, and suddenly you start getting a bad feeling. Something or someone makes you feel uneasy, and you've decided to pull out. Smart girl! Here's what that looks like:

$$(ab + ac) \quad \text{you, pulling out}$$

$$= a(b + c) \quad \text{of the party!}$$

In math, this is called **factoring,** and as you can see, it's distribution in reverse.

Ring Ring What's It Called?

Factoring is when we rewrite an expression as a **product of factors**. In other words, we rewrite an expression so that two or more things are now *multiplied times each other*. Factoring is distribution, in reverse! And to factor an expression *completely*, we must factor out the GCF—the greatest common factor.

· · · · · · · · · ·

* See p. 3 to review the definition of *term*. If you're not already comfortable with distribution problems like this one, check out Chapter 10 of *Kiss My Math*. We do lots of 'em there.

Factoring out the GCF

Here are some examples of factoring out the GCF—*undoing* distribution:

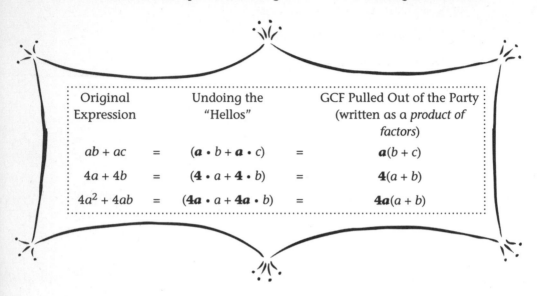

Original Expression		Undoing the "Hellos"		GCF Pulled Out of the Party (written as a *product of factors*)
$ab + ac$	=	$(\boldsymbol{a} \cdot b + \boldsymbol{a} \cdot c)$	=	$\boldsymbol{a}(b + c)$
$4a + 4b$	=	$(\boldsymbol{4} \cdot a + \boldsymbol{4} \cdot b)$	=	$\boldsymbol{4}(a + b)$
$4a^2 + 4ab$	=	$(\boldsymbol{4a} \cdot a + \boldsymbol{4a} \cdot b)$	=	$\boldsymbol{4a}(a + b)$

For each of the above equations, just multiply out the right sides, and you'll see that we get the left side in each case. Yep, they're equal to each other! The middle column isn't usually written out, but it shows us what's really going on.

Let's look at the last example from the above chart: $4a^2 + 4ab$. First, we need to figure out the GCF of both terms, which is $4a$.* So far, so good. Next, we'll rewrite each term so we can "see" the GCF, **4a**. And I like to draw parentheses around the expression, which reminds me how this is reverse distribution. It doesn't change the value of anything, so why not?

$$4a^2 + 4ab$$

$$= (\boldsymbol{4a} \cdot a + \boldsymbol{4a} \cdot b)$$

Finally, we pull out of the party:

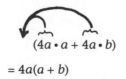

$$(4a \cdot a + 4a \cdot b)$$

$$= 4a(a + b)$$

. . . .

* To review finding the GCF with variables, see Chapter 2.

Interesting pattern. The guests who said hello to *everyone* are now the ones pulling out of this sketchy party. Must've been the weirdo in the back.

Step By Step

Factoring with variables—Reverse distribution:

Step 1. Use the Step By Step on p. 19 to find the GCF of the terms.

Step 2. Optional Step: Rewrite each term as a product of the GCF and whatever is left over so you can "see" the GCF, and draw parentheses around the whole thing.

Step 3. Time to pull the GCF out of the party! Use reverse distribution to rewrite the entire thing as a product of factors. The GCF should be one factor, and the other factor will be parentheses full of the "leftover" terms with addition and/or subtraction.

Step 4. Check your work by multiplying everything out using distribution, and make sure you get what you started with. This is especially important if you skipped Step 2. Done!

 QUICK NOTE   After we're done factoring, the leftover stuff in the parentheses should have no factors in common; in other words, they should be relatively prime. With $4a(a + b)$, since the leftover terms inside the parentheses, a and b, have no factors left in common, we know we've factored it completely.[*]

. . . .

[*] We could write the expression as a *product of factors* like this, $a(4a + 4b)$, but it wouldn't be completely factored until we also pulled out the 4.

Let's factor 3xy + 12y.

Step 1. First, we'll find the GCF of $3xy$ and $12y$, which we saw on p. 23 is **3y**.

Step 2. Now we'll rewrite the expression with the GCF pulled out of each term. And putting parentheses around the whole thing, our expression now looks like this: $(\mathbf{3y} \cdot x + \mathbf{3y} \cdot 4)$. It has the exact same value as $3xy + 12y$; it's just *written* differently.

Step 3. Doing reverse distribution, we'll pull the $3y$ to the outside: **3y(x + 4)**.

Step 4. Is that expression equal to what we started with, $3xy + 12y$? If we multiply it out, we'll see that yes, we did our reverse distribution (factoring) correctly. Check it out:

$$\mathbf{3y}(x + 4) \;=\; \mathbf{3y}(x) + \mathbf{3y}(4) \;=\; 3xy + 12y$$

Answer: 3y(x + 4)

📝 Doing the Math

Factor each expression by first figuring out the GCF of both terms, and then use reverse distribution to rewrite it as a product of factors. I'll do the first one for you.

1. $4e^2 - 5eh$

<u>Working out the solution</u>: Let's find the GCF of $4e^2$ and $5eh$. The coefficients, 4 and 5, are relatively prime; their GCF is 1. The only variable the terms share is e. Comparing e^2 to e, e wins! So the GCF is $1 \cdot e = e$. So we can rewrite $4e^2 - 5eh$ as $(e \cdot 4e - e \cdot 5h) = e(4e - 5h)$. Multiplying through, we can

check our answer: $e(4e - 5h) = e(4e) - e(5h) = 4e^2 - 5eh$.
Yep, we did it right!

Answer: $e(4e - 5h)$

2. $2ab + 10b^2$

3. $6c^2d - 7cd^2$

4. $7x + 2x$ (Hint: Follow the steps exactly, and then simplify your answer.)

(Answers on p. 402)

QUICK NOTE   So what's up with problem 4 above? I wanted to show you that a special case of factoring is *combining like terms!** So, for example, if we subtract $5x - 3x$, the answer is **2x** because 5 boxes minus 3 boxes is just 2 boxes, right? But also, because the GCF of $5x$ and $3x$ *is* **x**, and this is a free country after all, we could decide to factor out an x from the expression and get $5x - 3x = x(5 - 3) = x(2) = $ **2x**. Pretty nifty, huh? It's always good to see how these things are all connected. And you, my dear, have just gotten a whole lot smarter.

"Before I took algebra, I saw it as some sort of punishment! Now that I've gotten the hang of it, I never thought I would ever say this, but it can actually be really fun." **Dominique, 16**

. . . .

* See Chapter 9 in *Kiss My Math* to brush up on how to combine *like terms*. In a nutshell: When two terms have identical variable parts, you can add or subtract their coefficients.

Scary Expressions

"Boo!" Just kidding. Seriously, now that we have a feel for what's going on when we factor with reverse distribution like we're pulling out of the party, I'm going to show you my favorite way to factor. And it happens to be a lifesaver when the expressions get longer and scarier looking. Oh, and it involves cake. Birthday cake! Did I mention this was a birthday party?

Get this: The same Birthday Cake method from Chapter 2 can also be used to factor scary expressions! In fact, it finds the GCF *and* tells us what goes inside the parentheses. It gives us the entire answer, and it can be used for two or more terms. So it doesn't matter how long the expression is, the Birthday Cake method will work. Nice!

Let's factor a "scary" looking expression, and you'll see how birthday cake makes everything better.

QUICK NOTE   For each layer, it's important that we write in the + and – signs from the original expression. That way, we'll end up with the correct signs in the final answer.

Factor $18gh^3 + 12gh^2 - 6h^2$.

First, we draw a shelf underneath it: our first cake "layer." (See below.) Next, we look for a common factor shared by all three terms. How about **6**? So we can go ahead and factor out 6 and write it on the left. That will leave us with $3gh^3 + 2gh^2 - h^2$, which we'll write as our new layer directly below.

Hmm, does this newest layer have any common factors to all of its terms? Yep, h is in each term, so we draw another shelf layer. When we compare h^3, h^2, and h^2, we see that h^2 can safely be pulled out from each term, leaving us with $3gh + 2g - 1$, which we write along the bottom. Since the bottom layer has no common factors to *all* of its terms, we're done drawing layers.

$$6 \lfloor \overline{18gh^3 + 12gh^2 - 6h^2} \atop 3gh^3 + 2gh^2 - h^2} \rightarrow \begin{array}{l} 6\lfloor \overline{18gh^3 + 12gh^2 - 6h^2} \\ h^2 \lfloor \overline{3gh^3 + 2gh^2 - h^2} \\ \quad\; 3gh + 2g - 1 \end{array}$$

We'll multiply the terms on the left-hand side for the GCF: $6 \cdot h^2 = \mathbf{6h^2}$. And the leftover stuff along the bottom, $\mathbf{3gh + 2g - 1}$, goes in the parentheses. Our answer is $\mathbf{6h^2(3gh + 2g - 1)}$. And to make sure we did it correctly, let's multiply it all out using distribution:

$$6h^2(3gh + 2g - 1) \;=\; 6h^2 \cdot 3gh + 6h^2 \cdot 2g - 6h^2 \cdot 1 \;=\; 18gh^3 + 12gh^2 - 6h^2$$

And that's what we started with, so we did it right! Thanks, birthday cake.

Answer: $6h^2(3gh + 2g - 1)$

By the way, a problem doesn't *have* to be scary for the Birthday Cake method to work. (See problem 1 below.)

Doing the Math

Factor each expression using the Birthday Cake method. I'll do the first one for you.

1. $4e^2 - 5eh$

Working out the solution: This is the problem we did on p. 34 with reverse distribution; this time we'll use cake! Drawing a layer under it, we can write e on the left, and the new layer becomes $4e - 5h$. Since these two terms are relatively prime, this must be the last layer. The left side, e, is the GCF, and the bottom layer, $4e - 5h$, is what goes in the parentheses. (We already checked this answer on p. 35.)

$$e \;\big|\; \underline{4e^2 - 5eh} \\ \quad\;\; 4e - 5h$$

Answer: $e(4e - 5h)$

2. $15ab - 20b^2$

3. $12b^2cd - 18bcd$

4. $12b^2cd - 18bcd + 24bd$

5. $x^2y^2 + x^2y + xy^2 - xy$ *(Do this one step at a time. I know you can do it!)*

(Answers on p. 402)

⌐────────⌐

Factoring is like the Swiss army knife of math. It's essential for everything from reducing fractions to solving quadratic equations and more. It's one of the most powerful and omnipresent tools in math, and now it's in *your* tool belt.

⌐────────⌐

Takeaway Tips

 Factoring an expression is the reverse of the distributive property. Think of it this way: First you said hello to everyone at the party—that's the distribution—and then you "pulled out" of the party—that's the factoring!

To factor an expression completely, we must factor out the GCF.

To factor out the GCF, the Birthday Cake method works great! The numbers along the left side will multiply to become the GCF. And make sure to keep the addition and subtraction, if they exist, in each layer of the cake. That way, the bottom layer will end up being exactly what goes in the *parentheses* of the answer.

Danica's Diary
CONFESSIONALS:
ARE YOU BOY CRAZY?

Being a little boy crazy is normal. I mean, it's fun, and believe it or not, it's part of growing up. But if a guy's been on your mind a lot, ask yourself this: Is it a fun crush or a destructive obsession? Do you find yourself skipping schoolwork or dumbing yourself down for him? Or does he bring out the best in you? See what girls just like you are saying about it (and read the next page for my own story)!

"I dated this guy Erick for a whole year, and then he dumped me. After that I couldn't think of anything else. I just wanted to call him and tell him I loved him, and I wanted to know why he did it. But I knew he would think I was obsessed or something. I realize now that I <u>was</u> obsessed. I hardly did any of my schoolwork because I was daydreaming about Erick. It was eating me up inside, and I'm glad I finally stopped. Believe me, if you let all your crushes get to you, then you won't get to enjoy all the good stuff in life." **Michelle, 16**

"I daydream about guys sometimes, but I'm not one of those girls who has to have a boyfriend every second." **Daniella, 16**

"I have a huge crush on this one guy, and I think it makes me excel in volleyball; it's like I want to show off for him, and then I play better. I don't think being boy crazy is all bad." **Dawn, 15**

"I'm completely obsessed. It doesn't affect my schoolwork, but it still bugs me how one guy can drive me insane." **Ariana, 13**

"There was this guy Steve, who was first my friend, and then he asked to be my boyfriend, which was just amazing because I had a huge crush on him. But I started telling my mom I'd done my homework when I hadn't, just so I could hang out with him every day. I was getting obsessed. My grades went downhill, and we ended up breaking up anyway!" **Mae, 14**

"I'm so crazy about this one guy. My grades haven't gone down, but I end up blowing off my friends for him. I plow through people in the hallway just to be with him. It's pathetic! I lost a whole group of friends just because of him." **Sarah, 14**

"You can have crushes and all that stuff—just be careful who you go out with." **Candice, 15**

"Girls need to be strong and independent. It also helps you look less desperate if you are not thinking or trying to talk to a guy every second. Just focus on yourself and your friends—ironically, it will actually make guys like you more." **Lori, 16**

"In eighth grade, I liked a boy so much that I would get butterflies every time I saw him and I felt like I was going to puke. But then it seemed like I was changing everything about me just to try to get him to be with me. I realized that if a guy doesn't like you for who you are or expects you to change for him, he isn't worth it." **Brittany, 14**

"I've decided that when the perfect guy comes, I'll be waiting for him, but until then, I'm not out looking for a boyfriend. Once I reached this decision, it was a lot easier not to become obsessed over guys." **Tori, 14**

Listen, I struggled with being boy crazy, too. What I ended up learning was that I usually got obsessed with the *idea* of the guy more than the actual guy himself. I mean, how well do you REALLY know your crush? How can you be "in love" with someone you don't even know very well? See, our imaginations fill in the blanks, and of course our imaginations make the guy, well, nearly perfect. So of course we're going to want to be with our perfect imaginary boyfriend! And the real guy? Well, we use his *face* to daydream about, we seek out evidence to support "how great he is," and then we totally ignore the parts of the real guy that don't belong in our daydreams.

When I was 19, I did an acting job where I had to kiss this really cute guy. Let's call him "John." In

real life, he had a girlfriend, and I had a boyfriend.
But John was older and definitely a "bad boy" type—so
different from my boyfriend. I found myself thinking
about John all the time, and it freaked me out! Then
someone said to me, "You're obsessed with the *idea* of
John, not the real guy. You don't even really know
John." And of course that was exactly true. From then
on, I began to feel much better. I started thinking
about the not-so-perfect personality traits he
probably had, and I was able to get over those
feelings in just a few weeks. Phew!

If a guy is totally distracting you or making you
lovesick, take control of your feelings! Try some of
the ideas on pp. 113-15.

And by the way, if you ever feel tempted to
dumb yourself down for a guy, be sure to check out
pp. 182-83.

Meow Mix

Rational Expressions:
Fractions with Variables in Them

$\mathscr{F}$ractions with variables in them are called "rational expressions," which I've always found to be a rather ironic name, because half the time these fraction expressions look pretty crazy.* But the great news is that although they may *look* crazy, you already know how to deal with them. It's all the same stuff you've been doing with fractions for years; now we're just mixing in some variables. Some of you might remember getting help from our kitty-cat friends, the copycat fractions.† Well, these furry felines are back to help us with a whole mix of fraction déjà vu that will have you feeling like you're back in the fifth grade. Meow!

Multiplying/Dividing Fractions and Reciprocals

Just like with numbers, when we multiply fractions with variables, we can just multiply across the top and across the bottom, and we get our answer:

$$\text{Just Numbers: } \frac{2}{3} \cdot \frac{7}{5} = \frac{14}{15}$$

$$\text{With Variables: } \frac{2w}{y} \cdot \frac{x}{z} = \frac{2wx}{yz}$$

Not so bad, right? By the way, in this chapter, we're going to see a lot of variables in the denominators (bottoms) of fractions, so we'll assume the variables never have values that would make the denominator equal zero.

. . . .

* If you want to know the truth, we say <u>rational</u> because that's what a fraction is: a ratio.
† We talked about copycat fractions in Chapters 6–8 and 10 of *Math Doesn't Suck* for fractions with numbers. They'll say hi again in a few pages.

What's It Called?

The **reciprocal** of an expression is found by flipping its fraction upside down. If your expression is not already in fraction form, just put it over a 1 and then flip it. Everything on the top goes on the bottom, and everything on the bottom goes on the top. I like to say "re-FLIP-rocal"!

The reciprocal of $\frac{2}{3}$ is $\frac{3}{2}$.   The reciprocal of $\frac{5a}{b}$ is $\frac{b}{5a}$.

The reciprocal of $\frac{1}{3d}$ is $3d$.   The reciprocal of c $\left(\text{or } \frac{c}{1}\right)$ is $\frac{1}{c}$.

You may recall that in order to *divide* two fractions,* we just change the division symbol into multiplication, flip the second fraction, and then multiply. Great news: It works the exact same way with variables, so there's nothing new to learn!

flip it!

With Numbers: $\frac{1}{3} \div \frac{1}{2} = \frac{1}{3} \times \frac{2}{1} = \frac{1 \cdot 2}{3 \cdot 1} = \frac{2}{3}$

flip it!

With Variables: $\frac{1}{b} \div \frac{1}{c} = \frac{1}{b} \times \frac{c}{1} = \frac{1 \cdot c}{b \cdot 1} = \frac{c}{b}$

QUICK NOTE   A variable is simply a number whose value we don't know yet, right? So it makes sense that this fraction stuff should all pretty much work the same way with variables.

Let's practice!

. . . .

* To brush up on fraction division with numbers, see Chapter 5 in *Math Doesn't Suck*.

Doing the Math

Divide these fractions, just as if they were numbers. Assume all variables have nonzero values.* I'll do the first one for you.

1. $-\dfrac{x}{2} \div \dfrac{3y}{5x}$

<u>Working out the solution</u>: We just change the division into multiplication and FLIP the second fraction: $-\dfrac{x}{2} \times \dfrac{5x}{3y}$. Ta-da! And now we just multiply across the top and also across the bottom. We'll make sure to keep the negative sign, too:

$$-\dfrac{x \cdot 5x}{2 \cdot 3y} = -\dfrac{5x^2}{6y}.$$

Answer: $-\dfrac{5x^2}{6y}$

2. $\dfrac{a}{b} \div \dfrac{2}{c}$

3. $\dfrac{2}{c} \div \dfrac{a}{b}$

4. $\dfrac{3ef}{4} \div \dfrac{5}{e}$

5. $10 \div n$ (Hint: Rewrite 10 as $\dfrac{10}{1}$ and n as $\dfrac{n}{1}$.)

(Answers on p. 402)

Reducing Fractions with Variables

Reducing fractions with variables goes a little something like this:

$$\frac{3ab}{4a} = \frac{a \cdot 3b}{a \cdot 4} = \frac{a}{a} \cdot \frac{3b}{4} = 1 \cdot \frac{3b}{4} = \mathbf{\frac{3b}{4}}$$

After all, $\frac{a}{a}$ has the same thing on the top and bottom, so we know the fraction has to equal 1, right?[†] I like to call fractions with the same thing on top and bottom "copycat fractions." You can also "cancel" the a's, as

. . . .

* Just to make sure we never divide by zero!
† Unless of course, $a = 0$, in which case $\frac{a}{a}$ and $\frac{3ab}{4a}$ are undefined.

long as you're careful.* Just be sure you don't try to cancel when addition or subtraction is involved like this:

$$\frac{3a + b}{4a}$$

There's simply nothing we can cancel from the top and bottom in the above fraction—there are no common factors. Oh well, better luck next time.

Hidden Copycats

To escape detection, cats have a way of disappearing into the night . . . and hiding in paper bags. Very crafty. Check out the fraction below. Even though it involves addition and subtraction, there *is* a way to reduce it . . . if we can find the hiding copycat.

$$\frac{3ab + 5a}{4a - ac}$$

Remember how we factored expressions with reverse distribution in Chapter 3? We're about to use that little skill again. See, sometimes the top and bottom of a fraction will have a common factor that can be pulled out. And if those common factors match each other, then we've just found a hidden copycat!

$$\frac{3ab + 5a}{4a - ac} = \frac{\mathbf{a}(3b + 5)}{\mathbf{a}(4 - c)} = \frac{\mathbf{a}}{\mathbf{a}} \cdot \frac{(3b + 5)}{(4 - c)} = 1 \cdot \frac{(3b + 5)}{(4 - c)} = \frac{\mathbf{3b + 5}}{\mathbf{4 - c}}$$

Ta-da! It's important to remember that we haven't changed the *value* of the fraction; we've just rewritten it, exactly in the same way that we might rewrite $\frac{3}{6}$ as $\frac{1}{2}$. It just looks a bit more complicated because of the variables.

Step By Step

Reducing fractions with variables:

Step 1. Completely factor the top and bottom of the fraction, so that each is written as a product of factors, using reverse distribution as we learned in Chapter 3.

· · · · · · · · · ·

* For more practice canceling variable factors, see Chapter 8 in *Kiss My Math*.

Step 2. Look for hiding copycats! If there are any common factors on top and bottom, rewrite the fraction as two separate fractions multiplying times each other, where one of them is a copycat fraction.

Step 3. Because the copycat fraction, which equals 1, is *multiplied* times something else, it can now "poof" go away (this is what canceling really is). And if there are no more hiding copycats, AKA common factors on top and bottom, then you're done!

Watch Out!

Copycat fractions equal 1; this means they can "poof" go away when they are *multiplied* times something else—*not* when they are added or subtracted on their own. For example:

$$\frac{x}{x} + 3x \neq 3x \qquad \text{But this is correct: } \frac{x}{x} + 3x = \mathbf{1 + 3x}$$

QUICK NOTE   We must assume in all these problems that the denominators never equal zero.[*] I know I keep bringing that up, but it's worth repeating. After all, we can never, ever divide by zero. It's simply not defined in the rules of this little game we play called "math."

And... Action! Step By Step In Action

Reduce this fraction: $\dfrac{m}{2mn - 2m}$.

Step 1. There's nothing to factor on the top, so let's factor the bottom. The GCF of $2mn$ and $2m$ is **2m**, so we can factor it out and rewrite the

. . . .

[*] For example, if the fraction were $\dfrac{x}{2x}$, we would assume that $2x \neq 0$; if the fraction were $\dfrac{1}{a + b}$, we would assume that $(a + b) \neq 0$, and so on.

bottom as $2m(n-1)$. Our fraction now looks like $\dfrac{m}{2m(n-1)}$. So far, so good?

Step 2. It's time to rewrite $\dfrac{m}{2m(n-1)}$ as two fractions multiplying together. The only factor the top and bottom share is m, so the copycat fraction will be $\dfrac{m}{m}$.

On top, we can stick in a factor of 1 to make things clearer. Those kitties aren't the only clever ones.

$$\frac{m}{2m(n-1)} = \frac{m \cdot 1}{m \cdot 2(n-1)} = \frac{m}{m} \cdot \frac{1}{2(n-1)}$$

Step 3. Now our copycat fraction, which equals 1, can "poof" go away, and we're done!

$$\frac{m}{m} \cdot \frac{1}{2(n-1)} = 1 \cdot \frac{1}{2(n-1)} = \frac{1}{2(n-1)}$$

We could also distribute that 2 if we wanted: $\dfrac{1}{2n-2}$. Both answers are perfectly fine.*

Answer: $\dfrac{1}{2(n-1)}$ or $\dfrac{1}{2n-2}$

"I try to catch myself before I say anything 'dumb on purpose' around a boy, because I realize it just isn't attractive, and it's not worth it to seem dumb! Plus it doesn't feel very good." **Rebecca, 15**

Take Two: Another Example

Reduce this fraction: $\dfrac{3xy + 12y}{x + 4}$.

Step 1. Hmm, looking at the bottom of the fraction, there is no GCF that can be pulled out. So at first it might seem like nothing can be done to

. . . .

* I prefer the first one, but your teacher might prefer that you leave your answer factored (like the first one) or distributed (like the second one), so be sure to check.

reduce this fraction, right? But let's have a little faith, factor the top, and see what happens. On p. 34, we factored $3xy + 12y$ and rewrote it as $3y(x + 4)$, which means we can rewrite our fraction like this, and look what happens!

$$\frac{3xy + 12y}{x + 4} = \frac{3y(x + 4)}{x + 4}$$

Steps 2 and 3. Interesting, it seems we have a common factor after all: $(x + 4)$. A crafty kitty, indeed. So this means our copycat will be $\frac{(x + 4)}{(x + 4)}$. It's time to split things up! Notice we'll stick in a factor of 1 to keep things clear. FYI, if we just "canceled" $(x + 4)$ from the top and bottom, we'd still end up with that 1 on the bottom.

$$\frac{3y(x + 4)}{x + 4} = \frac{3y \cdot (x + 4)}{1 \cdot (x + 4)} = \frac{3y}{1} \cdot \frac{(x + 4)}{(x + 4)} = \frac{3y}{1} \cdot 1 = 3y$$

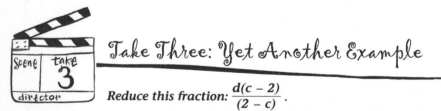

Answer: $3y$

QUICK NOTE   Here's a little catnip trick: You know that $(-1)(-1) = 1$, right?* Sometimes it will be to our advantage to multiply $(-1)(-1)$ times something, and then only distribute *one* of the -1's. Sometimes this can coax out a stubborn copycat. See below for an example.

Take Three: Yet Another Example

Scene | take **3** | director

Reduce this fraction: $\frac{d(c - 2)}{(2 - c)}$.

Steps 1 and 2. So close to having a copycat, yet so far away . . . But wait, there's something we can do! Go with me here for a second: Let's multiply the $(2 - c)$ on the bottom by $(-1)(-1)$, which is the same as

• • • • • • • • • •

* To review multiplication of negative numbers, see Chapter 3 in *Kiss My Math*.

multiplying it by 1, so we're not changing its value at all. The key is that we'll only distribute *one* of the –1's:

$$(2 - c) = (-1)(-1)(2 - c) = (-1)(-2 + c) = (-1)(c - 2)^*$$

Now we can just replace the bottom of our fraction, $(2 - c)$, with $(-1)(c - 2)$, right? Totally legal, since we've just seen that those two expressions are *equal* to each other. So:

$$\frac{d(c - 2)}{(2 - c)} = \frac{d(c - 2)}{(-1)(c - 2)} = \frac{d}{-1} \cdot \frac{(c - 2)}{(c - 2)} \text{ (copycat, yay!)} \frac{d}{-1} \cdot 1 = -d$$

Step 3. We've outwitted the crafty kitty and discovered that $\frac{d(c - 2)}{(2 - c)} = -d$.[†]

Answer: –*d*

QUICK NOTE   It doesn't matter if you multiply the top or the bottom by (–1)(–1); you'll get the same final answer either way. Try it with the above example and you'll see for yourself!

Doing the Math

Rewrite these fractions into reduced form. Assume no denominators equal zero. I'll do the first one for you.

1. $\dfrac{7x - 7y}{y - x}$

. . . .

* Do you see why $(-2 + c) = (c - 2)$? Check it out: $-2 + c = c + (-2) = c - 2$.
† Unless of course $c = 2$, which isn't allowed, because then we'd have a denominator of zero. And that would mean our first expression wasn't defined in the first place! How embarrassing.

<u>Working out the solution</u>: Factoring 7 out of the top with reverse distribution, we can rewrite our fraction $\dfrac{7x - 7y}{y - x}$ as $\dfrac{7(x - y)}{y - x}$. Now let's be clever and multiply $(-1)(-1)$ times the top, only distributing one of the -1's, and we get $\dfrac{7(-1)(-1)(x - y)}{y - x} = \dfrac{7(-1)(-x + y)}{y - x} = \dfrac{7(-1)(y - x)}{y - x}$. Now we have *identical* factors on the top and bottom. I see a kitty cat! $\dfrac{7(-1)}{1} \cdot \dfrac{y - x}{y - x} = \dfrac{-7}{1} \cdot 1 = -7$.

Answer: −7

2. $\dfrac{9ab - 3b}{6b + 3bc}$

3. $\dfrac{15xy + 5x}{5x}$

4. $\dfrac{11cd - 11d}{7c - 7}$

5. $\dfrac{8a - 4b}{b - 2a}$ *(Hint: Multiply the bottom by (−1)(−1), and only distribute one of the −1's.)*

(Answers on p. 402)

"When I was younger, I wanted to be a writer, and now I want to be an ER doctor! We change our minds so much. If we study in all of our classes, I believe it will help us all in the long run." **Chloe, 13**

Complex Fractions—with Variables!

Here are two complex fractions:* one with just numbers, and one with variables.

. . . .

* To review complex fractions with just numbers, see Chapter 9 of *Math Doesn't Suck*.

$$\frac{\frac{1}{6}}{\frac{3}{4}} \quad \text{and} \quad \frac{\frac{x}{y}}{\frac{2x}{3}}$$

As you know, a fraction *is* a division problem: "top ÷ bottom" (for example, $\frac{10}{2} = 10 \div 2$ and $\frac{3}{4} = 3 \div 4$). In fact, it doesn't matter what is on the top and bottom—more fractions, a flower, a fish,* whatever—<u>a fraction means division</u>. 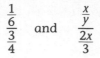 $= \text{🕊} \div \text{🌸}$

So we can rewrite complex fractions as fraction division, which we totally know how to handle from p. 43. Then we'll seek out hiding copycat fractions, if any exist, and reduce the final answer:

flip it!

With Numbers: $\dfrac{\frac{1}{6}}{\frac{3}{4}} = \dfrac{1}{6} \div \dfrac{3}{4} = \dfrac{1}{6} \times \dfrac{\mathbf{4}}{\mathbf{3}} = \dfrac{4}{18} = \dfrac{\mathbf{2}}{\mathbf{2}} \cdot \dfrac{\mathbf{2}}{\mathbf{9}} = \dfrac{\mathbf{2}}{\mathbf{9}}$

flip it!

With Variables: $\dfrac{\frac{x}{y}}{\frac{2x}{3}} = \dfrac{x}{y} \div \dfrac{2x}{3} = \dfrac{x}{y} \times \dfrac{\mathbf{3}}{\mathbf{2x}} = \dfrac{3x}{2xy} = \dfrac{\mathbf{x}}{\mathbf{x}} \cdot \dfrac{3}{2y} = \dfrac{\mathbf{3}}{\mathbf{2y}}$

Step By Step

Simplifying complex fractions by rewriting them as division:

Step 1. Rewrite the complex fraction as a division problem, with the top fraction divided by the bottom fraction.

Step 2. Using normal fraction division techniques, change the division symbol to multiplication and flip the second fraction (in other words, take its reciprocal).

Step 3. Multiply the fractions, and reduce if possible. Done!

. . . .

* A request from the copycat. 🐱

QUICK NOTE If any of the terms involve addition or subtraction, be sure to use parentheses to keep everything straight.

And... Action! Step By Step In Action

Simplify $\dfrac{\dfrac{x}{y+1}}{\dfrac{4}{5}}$.

Step 1. Rewriting it as the top divided by the bottom, we get $\dfrac{x}{(y+1)} \div \dfrac{4}{5}$. We put parentheses around the $(y+1)$, just to help keep everything straight.

Steps 2 and 3. Now let's rewrite the division as usual, with multiplication and flipping the second fraction: $\dfrac{x}{(y+1)} \times \dfrac{5}{4} = \dfrac{5x}{4(y+1)}$. There are no common factors (no hiding copycats!) so we're done.

Answer: $\dfrac{5x}{4(y+1)}$ **or** $\dfrac{5x}{4y+4}$

Shortcut Alert

On p. 97 in *Math Doesn't Suck*, I taught you a shortcut for complex fractions called "means and extremes," which we can also use when variables are involved.

$$\text{means} \left[\dfrac{\dfrac{x \leftarrow}{y+1}}{\dfrac{4}{5} \leftarrow} \right] \text{extremes} = \dfrac{5x}{4(y+1)}$$

The two "extreme" terms, x and 5, multiply together to give us the numerator of our new fraction, and the two middle (mean) terms, 4 and ($y + 1$), multiply together to give us our new denominator. Ta-da!

This shortcut saves a few steps, but we're really doing the same thing as the division method, only now the flipping and multiplication happen all at once.

Watch Out!

In order to use the above shortcut, <u>we must have one fraction on top of another fraction</u>. So if you see something like this, $\dfrac{\frac{x}{2}}{6x}$, it's not ready for the shortcut. But since $6x = \dfrac{6x}{1}$, we can just rewrite our problem like this, $\dfrac{\frac{x}{2}}{\frac{6x}{1}}$, and proceed as normal!

QUICK (REMINDER) NOTE* Like we saw on p. 6, multiplying something times a fraction is the same as multiplying that same thing times the numerator and keeping the denominator. So, $\frac{n}{2}(5) = \frac{5n}{2}$, and $\frac{1}{2}x = \frac{x}{2}$. This can be useful in complex fractions.

For example, $\dfrac{\frac{1}{2}x}{3}$ is not ready for the shortcut, but if we rewrite

it like this, $\dfrac{\frac{x}{2}}{\frac{3}{1}}$, now it is ready.

. . . .

* For more on these kinds of fraction properties, check out pp. 115–18 in *Kiss My Math*.

Doing the Math

Simplify these with the step-by-step method on p. 51 or by using the shortcut on p. 52. I'll do the first one for you.

1. $\dfrac{ab}{\frac{1}{2}a}$

<u>Working out the solution</u>: To use the shortcut, we'll have to put the numerator and denominator each into fraction form: $\dfrac{\frac{ab}{1}}{\frac{a}{2}}$

and then using means and extremes, this becomes $\dfrac{2ab}{1a}$. Now let's factor out that hiding kitty cat and reduce: $\dfrac{a}{a} \cdot \dfrac{2b}{1} = \textbf{2b}$. Done!

Answer: 2b

2. $\dfrac{\frac{x}{y}}{\frac{x}{4}}$

3. $\dfrac{\frac{a}{b}}{\frac{a}{b}}$

4. $\dfrac{x+1}{\frac{3}{2}}$

(Answers on p. 403)

Takeaway Tips

The reciprocal is found by flipping a fraction upside down: Re-FLIP-rocal!

To reduce fractions with variables in them, first do reverse distribution on the top and bottom, and then factor out the hiding copycat fractions.

Sometimes multiplying $(-1)(-1)$ times the top *or* bottom of a fraction can coax out a hiding copycat and allow you to simplify the fraction.

 To simplify complex fractions, either write them as the division problems they really are, or use the means and extremes shortcut. Remember, to use the shortcut, the numerator and denominator must both be written as fractions.

 Variables can never be anything that makes the denominator of a fraction equal zero.

 Remember to use parentheses whenever addition or subtraction is involved; it'll prevent many mistakes!

hold me back in life. Ironically, math is the subject that helped me with my confidence the most. In math, an answer is either right or wrong, and I found that as my math improved, I gained more confidence and was able to speak up in class. Sometimes I had to force myself to, but I kept participating. Math was the only thing that really let me prove to myself that I could answer in class without judgment or ridicule, and that it is okay to be wrong sometimes. But even though math helped me become assertive, I had no idea it would be central to my career.

But I took a class in TV production my senior year, and I was instantly hooked. When I learned that math could take me far in this field, I was thrilled! Today, I sell commercial airtime for TV stations across the country, so I get to travel to visit clients and entertain them at lunches, dinners, sporting events, and the occasional TV show taping—fun! And, of course, I also get to flex my math muscles daily: Math is essential to predicting what the airtime is worth before I can negotiate a deal. The more people who watch a show, the more that time block is worth to an advertiser who wants lots of people to see their commercial. Makes sense, right? But since TV-watching habits fluctuate, there is no way of knowing how many people are going to watch a show (I can't predict the future!), so I study past research—charts, numbers, and formulas—and make my own estimate about what I think the viewing habits will be in the future for a particular group of shows. That way, I can explain to the advertisers how much I think they should pay for the airtime and make the deal.

In literal terms, I sell airtime, but what I'm really selling is math—and I couldn't do my job without being able to speak up and offer my carefully calculated answer...even if there's a chance that my estimate will turn out to be wrong. Getting good at math has given me that confidence and has made me extremely successful at what I do. Thanks to the girl who learned to love math and to speak up in class, I get to work in an exciting field with interesting people—and dress up in cute pencil skirts and pumps while doing it!

Chapter 5

Eating Pizza in the Dark

Adding and Subtracting Fractions with Variables

$\mathcal{H}$ave you ever eaten pizza in the dark? Talk about a surefire way to stain your brand new white pants. Adding and subtracting fractions with variables is a lot like eating pizza in the dark (except for the pants-staining part).

Let's say you're having two friends over for a pizza party. The three of you are going to watch movies in the dark and eat two small pizzas, one pepperoni pizza and one veggie pizza. Fun! You have to share with your older brother, too, so that's 4 people total, which means each person should get a quarter of *each* pizza. One of your friends is vegetarian, but as long as each person gets a total of a half pizza, things will be totally fair, right?

These small pizzas come divided into 6 pieces, and before you get a chance to recut them, your brother runs into the kitchen, saying, "Mom says I get two pieces of each!" Then he grabs 2 of the pieces of the pepperoni pizza! He's just taken $\frac{2}{6}$ of the pepperoni pizza, in other words, $\frac{1}{3}$ of it. Any minute he'll run back in and grab two pieces of veggie pizza, which would give him more than his fair share! Luckily, a little math goes a long way toward outsmarting older brothers. We'll deal with him in a moment.

For now, let's go back to the basics of adding fractions. Remember that when we have the same denominator, we can just add the numerators, because it's like adding the same-sized pizza slices together.

$$\frac{1}{8} + \frac{2}{8} = \frac{3}{8}:$$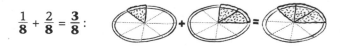

Well, the same is true when there are variables in fractions. Remember, variables are numbers; we just don't happen to know *which* numbers they are yet. So this is true:

$$\frac{1}{w} + \frac{2}{w} = \frac{3}{w}$$

Think about that for a moment: In this case, it might be pitch black and we might not know what size pizza pieces we're talking about (in this case, the pizza was cut into *w* pieces), but the same rules work; as long as we have a **common denominator**, we can add the fractions by adding across the top.

What's It Called?

For two or more fractions to have a **common denominator**, they must have the *exact same* denominator—no matter how crazy it might look. For example, these pairs of fractions each have common denominators:

$$\frac{1}{2} \text{ and } \frac{5}{2} \qquad \frac{n}{4n} \text{ and } \frac{4}{4n} \qquad \frac{5x}{y(y+1)} \text{ and } \frac{y}{y(y+1)}$$

I guess if we were eating pizza in the dark, it would be harder to keep track of how *many* pieces we were eating, so we can end up with variables in the numerator, too:

$$\frac{x}{w} + \frac{y}{w} = \frac{x+y}{w}$$

Again, just because we don't know much about the pizza in front of us doesn't mean it doesn't follow all the typical pizza rules. So with a *common denominator*—no matter what it is—we can simply add or subtract across the top, and the answer will keep that same denominator.*

. . . .

* For the rest of this chapter, assume that no variable has a value that could make a denominator zero.

Doing the Math

Add or subtract these fractions with common denominators. I'll do the first one for you.

1. $\dfrac{6x}{4} - \dfrac{x+y}{4}$

<u>Working out the solution</u>: No matter how tempted we might be to reduce that first fraction, we better not! We need that denominator to be the same as the one in the second fraction; this allows us to simply subtract across the top: $\dfrac{6x - (x+y)}{4}$. Make sense? The parentheses help us keep track of that negative sign, which we now distribute: $\dfrac{6x - x - y}{4}$. Now we just subtract $6x - x$ on the top, and we get $\dfrac{5x - y}{4}$. Done!

Answer: $\dfrac{5x - y}{4}$

2. $\dfrac{x}{5} + \dfrac{2x}{5}$

3. $\dfrac{x}{y} + \dfrac{2x}{y}$

4. $\dfrac{1}{b} - \dfrac{a+1}{b}$

(Answers on p. 403)

Watch Out!

Be sure to distribute negative signs when subtracting fractions, as in problem 1 and problem 4 above. For example, if we didn't distribute, we could have gotten $\dfrac{6x - (x+y)}{4} \rightarrow \dfrac{6x - x + y}{4} \rightarrow \dfrac{5x + y}{4}$. Oops! This is one of those times when it's especially helpful not to skip steps and to remember that parentheses are our *friends*. Use 'em to remind yourself of who is grouped together.

Adding and Subtracting Fractions
with Different Denominators

Before we can add or subtract fractions with different denominators, just like we've always done with numbers, we need to rewrite one or both fractions using a copycat fraction* so that they *do* have the same denominator. In both of the cases below, we'll multiply the first fraction by the copycat $\frac{3}{3}$ in order to get matching denominators.

With Numbers:

$$\frac{1}{2} + \frac{5}{6} \;=\; \frac{3}{3} \cdot \frac{1}{2} + \frac{5}{6} \;=\; \frac{3}{6} + \frac{5}{6} \;=\; \frac{3+5}{6} \;=\; \frac{8}{6} \;=\; \frac{4}{3}$$

With Variables:

$$\frac{a}{2} + \frac{5a}{6} \;=\; \frac{3}{3} \cdot \frac{a}{2} + \frac{5a}{6} \;=\; \frac{3a}{6} + \frac{5a}{6} \;=\; \frac{3a+5a}{6} \;=\; \frac{8a}{6} \;=\; \frac{4a}{3}$$

Even with variables on the bottom, like with $\frac{1}{x} + \frac{5}{xy}$, we can still come up with a copycat fraction (in this case $\frac{y}{y}$) that allows the fractions to be rewritten so they have a common denominator. We just pick whatever copycat we need in order to get the denominators to match. Remember, all copycats equal 1, so they aren't changing anyone's value:[†]

$$\frac{1}{x} + \frac{5}{xy} \;=\; \frac{y}{y} \cdot \frac{1}{x} + \frac{5}{xy} \;=\; \frac{y}{xy} + \frac{5}{xy} \;=\; \frac{y+5}{xy}$$

copycat↗   same denominator, yay!↖↗

QUICK NOTE If you ever get confused about how to do something with fractions and variables, try doing a similar problem with just numbers. It'll get your brain back on straight, and you'll feel ready to tackle the actual problem in front of you. It really works!

· · · · · · · · · ·

* I explain who the copycat fractions are in Chapter 4; these are fractions that equal 1 because their numerators and denominators are identical.
† Remember, variables *aren't allowed* to have values that make a denominator equal zero. In higher math, the answer would look like this: $\frac{y+5}{xy}$, $x, y \neq 0$. So strict, I know!

Lowest Common Denominator

When it comes to pizza, big is good. When it comes to math, small is much better. When we need a common denominator, the smallest possible option is called the **lowest common denominator**.

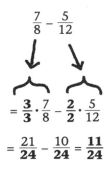

Ring Ring What's It Called?

The **lowest common denominator** (LCD) for two or more fractions is just another way of saying the *lowest common multiple* (LCM) of the denominators. For example:

Because the LCM of 6 and 15 is **30**, the LCD of $\frac{1}{6}$ and $\frac{1}{15}$ is **30**.

Because the LCM of 5x and xy is **5xy**, the LCD of $\frac{1}{5x}$ and $\frac{1}{xy}$ is **5xy**.

Just like we saw with LCMs on p. 22, the LCD will be the smallest (simplest) expression that has both original denominators as *factors*.

With a problem like $\frac{7}{8} - \frac{5}{12}$, we don't really want to use the denominator 8 × 12 = **96**, do we? Um, no, thank you very much. We'd rather use a smaller common denominator—in fact, we want the smallest number that has both 8 and 12 as factors, and that is **24**.* Using the right copycat fractions, we can rewrite our problem with a common denominator of 24 and then just subtract across the top:

$$\frac{7}{8} - \frac{5}{12}$$

$$= \frac{\mathbf{3}}{\mathbf{3}} \cdot \frac{7}{8} - \frac{\mathbf{2}}{\mathbf{2}} \cdot \frac{5}{12}$$

$$= \frac{21}{\mathbf{24}} - \frac{10}{\mathbf{24}} = \frac{\mathbf{11}}{\mathbf{24}}$$

* To review finding LCDs of fractions with numbers, see Chapter 8 in *Math Doesn't Suck*.

And we can do the same thing when variables are involved. What's the LCD for $\frac{1}{2x} + \frac{1}{xy}$? Well, the smallest expression that has both $2x$ and xy as factors would be **2xy**, right? So we'd rewrite the fractions with a denominator of $2xy$, using the appropriate copycat fractions to make this happen. In this case, we'll need to use the copycats $\frac{y}{y}$ and $\frac{2}{2}$.

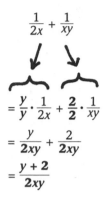

$$\frac{1}{2x} + \frac{1}{xy}$$

$$= \frac{y}{y} \cdot \frac{1}{2x} + \frac{2}{2} \cdot \frac{1}{xy}$$

$$= \frac{y}{2xy} + \frac{2}{2xy}$$

$$= \frac{y + 2}{2xy}$$

See, it doesn't matter if there are variables on the top or bottom of the fractions (or even both!), the method is the same as with numbers: We figure out what common denominator we want, and then we use copycat fractions to rewrite the fractions so they have that denominator. Then we're allowed to add or subtract across the top. Nice.

Step By Step

Adding/subtracting fractions with different denominators:

Step 1. Find the LCM of the denominators (the LCD): That will be our new denominator for both fractions. You can either eyeball them or use the Birthday Cake method from p. 23.

Step 2. Determine *which copycat fractions* you need to multiply to each fraction so they can get this new denominator, and then do the multiplication!

Step 3. Now that the fractions have been rewritten so they have the same denominator, just add/subtract across the top, keeping the same denominator. Reduce if possible. Done!

QUICK NOTE   Sometimes the LCD will just be the two denominators multiplied together. This will happen whenever the original denominators didn't have any factors in common. Oh well.

And... *Action!* Step By Step In Action

Add $\frac{1}{x} + \frac{1}{x+1}$.

Step 1. These denominators have no factors in common, so the LCD will just be their product: $x(x+1)$.

Step 2. Now we'll use the necessary copycat fractions to write each fraction with this denominator. In order to attain $x(x+1)$ as the denominator, notice that $\frac{1}{x}$ will need the copycat $\frac{(x+1)}{(x+1)}$ and that $\frac{1}{(x+1)}$ will need the copycat $\frac{x}{x}$. Rewriting our problem:

$$\frac{1}{x} + \frac{1}{x+1} = \frac{(x+1)}{(x+1)} \cdot \frac{1}{x} + \frac{x}{x} \cdot \frac{1}{(x+1)} = \frac{(x+1)}{x(x+1)} + \frac{x}{x(x+1)}$$

copycats   same denominator, yay!

Step 3. Adding across the top, we get $\frac{(x+1)+x}{x(x+1)} = \frac{2x+1}{x(x+1)}$.

There are no hiding copycats—the bottom is already factored, and the top can't be factored (the GCF of $2x$ and 1 is 1), so we know there is no way to reduce this further. If we wanted, we could distribute that x on the bottom; it doesn't really matter.

Answer: $\frac{2x+1}{x(x+1)}$ **or** $\frac{2x+1}{x^2+x}$

Hmm, kind of messy, but at least we know how to do it.

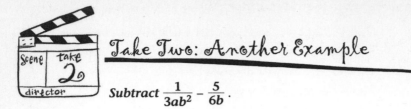
Subtract $\dfrac{1}{3ab^2} - \dfrac{5}{6b}$.

Step 1. We already found the LCM of $3ab^2$ and $6b$ on p. 24 using the Birthday Cake method. Their LCM is **$6ab^2$**, which will be our common denominator: our LCD.

Step 2. Now we'll use the copycat fractions we need, in this case $\dfrac{2}{2}$ and $\dfrac{ab}{ab}$, to rewrite our problem so it has this common denominator:

$$\frac{2}{2} \cdot \frac{1}{3ab^2} - \frac{ab}{ab} \cdot \frac{5}{6b} \;=\; \frac{2}{6ab^2} - \frac{5ab}{6ab^2}$$

Step 3. Now we can just subtract across, and we get $\dfrac{2 - 5ab}{6ab^2}$. Can we reduce this? The top can't be factored at all (the GCF of 2 and $5ab$ is 1), and the bottom certainly doesn't have the factor $2 - 5ab$, so that means there are no hiding copycats. We're done!

Answer: $\dfrac{2 - 5ab}{6ab^2}$

QUICK NOTE   To add a non-fraction to a fraction like this, $x + \dfrac{1}{5}$, first write both as fractions: $\dfrac{x}{1} + \dfrac{1}{5}$. In this case, the LCD is 5, so only the first fraction will need a copycat:

$$\frac{x}{1} + \frac{1}{5} \;=\; \frac{5}{5} \cdot \frac{x}{1} + \frac{1}{5} \;=\; \frac{5x}{5} + \frac{1}{5} \;=\; \frac{5x + 1}{5}.$$

Voilà! We've successfully added them together.

"Algebra isn't as hard as I thought it was going to be. There are still times that I don't understand it, but I just keep trying." **Amanda, 14**

Doing the Math

Combine the following expressions, and reduce if possible. Assume no denominators equal zero. I'll do the first one for you.

1. $\dfrac{1}{3ab^2} - \dfrac{1}{6b} + \dfrac{2c}{a^2}$

<u>Working out the solution</u>: So, we need the LCM of $3ab^2$, $6b$, and a^2.* On p. 24, we found that the LCM of $3ab^2$ and $6b$ is $6ab^2$. Next, we need the LCM of $6ab^2$ and the remaining denominator, a^2. We could use the Cake method or just eyeball it, and we get **$6a^2b^2$**. Using various copycats to rewrite our problem with **$6a^2b^2$** on the bottom, we get:

$$\frac{2a}{2a} \cdot \frac{1}{3ab^2} - \frac{a^2b}{a^2b} \cdot \frac{1}{6b} + \frac{6b^2}{6b^2} \cdot \frac{2c}{a^2} = \frac{2a}{6a^2b^2} - \frac{a^2b}{6a^2b^2} + \frac{12b^2c}{6a^2b^2}$$

And now we just combine across the top. There's no way to reduce, so we're done!

Answer: $\dfrac{2a - a^2b + 12b^2c}{6a^2b^2}$

2. $\dfrac{1}{2a} + \dfrac{3}{a}$

3. $\dfrac{1}{3} + \dfrac{2}{v}$

4. $\dfrac{1}{n} + \dfrac{1}{1-n}$

5. $x - \dfrac{1}{y}$ *(Hint: Write x as the fraction $\frac{x}{1}$.)*

6. $\dfrac{a}{6b} + \dfrac{1}{2a} + \dfrac{c}{4ab}$

7. $\dfrac{7}{2(y-1)} + \dfrac{1}{2y}$ *(Take it one step at a time: You can do it!)*

(Answers on p. 403)

· · · · · · · · ·

* To review finding LCMs for three expressions, see p. 25.

It's time to mix together everything that we've done in the last few chapters . . .

Crazy Rational Expressions? No Problem!

"Crazy rational" might sound like an oxymoron, but don't worry! You can totally handle wacky-looking expressions that might have baffled you in the past.

For example, let's simplify $\dfrac{1}{\frac{1}{x} - \frac{1}{y}}$.

Hmm, this looks a little like the complex fractions that we dealt with in Chapter 4, except that now we've also got *fraction subtraction* in the denominator. Let's break this into smaller, bite-sized pieces. We'll pretend nothing else exists in the world except this denominator, and do just that subtraction. The LCM of x and y is xy, so our LCD is xy, and we get $\dfrac{1}{x} - \dfrac{1}{y} = \dfrac{y}{y} \cdot \dfrac{1}{x} - \dfrac{x}{x} \cdot \dfrac{1}{y} = \dfrac{y}{xy} - \dfrac{x}{xy} = \dfrac{y - x}{xy}$. That's our new denominator, and our entire problem now looks like this: $\dfrac{1}{\frac{y - x}{xy}}$.

You might not think that looks much simpler than how it started, but the denominator is now a *single fraction*, so we've made progress. Now we can treat it like the complex fractions we simplified on p. 51. At this point, we have two choices. We can either write this as the division problem it really is:

flip it!

$$\dfrac{1}{\frac{y - x}{xy}} = 1 \div \dfrac{y - x}{xy} = 1 \times \dfrac{xy}{y - x} = \dfrac{xy}{y - x}$$

. . . or we could use the "means and extremes" shortcut from p. 52 and get the same answer. We just need to write the numerator of 1 as $\frac{1}{1}$, and we get:

$$\dfrac{1}{\frac{y - x}{xy}} = \left(\dfrac{\frac{1}{1}}{\frac{y - x}{xy}}\right) = \dfrac{1 \cdot xy}{1 \cdot (y - x)} = \dfrac{xy}{y - x}$$

There's nothing to reduce, so we're done!

Answer: $\dfrac{xy}{y - x}$

It seems kinda strange that $\frac{xy}{y-x}$ has the same value as $\frac{1}{\frac{1}{x}-\frac{1}{y}}$. But just plug a few numbers in (like $x = 4$ and $y = 2$) and you'll see that it's really true.*

Step By Step

Simplifying complicated rational expressions:

Step 1. Simplify the numerator and denominator separately, as if they live in their own separate worlds.

Step 2. Make sure you now have a complex fraction with just *a single fraction on top of another single fraction*. Next, simplify it by either rewriting it as a division problem (see p. 51) or by using the means and extremes shortcut from p. 52.

Step 3. Now that you have a single fraction, reduce if possible. Use reverse distribution and factor out any hidden copycat fractions.

And... Action! Step By Step In Action

Simplify $\dfrac{\frac{3x}{5}}{\frac{x}{4}+\frac{x}{2}}$.

Step 1. Let's deal with the denominator. To add $\frac{x}{4}+\frac{x}{2}$, we'll need an LCD of 4, so we'll use the copycat fraction $\frac{2}{2}$:

$$\frac{x}{4}+\frac{x}{2} \ = \ \frac{x}{4}+\frac{2}{2}\cdot\frac{x}{2} \ = \ \frac{x}{4}+\frac{2x}{4} \ = \ \frac{3x}{4}$$

Step 2. Plugging in our simplified denominator, $\frac{3x}{4}$, our problem now becomes $\dfrac{\frac{3x}{5}}{\frac{3x}{4}}$. Using the means and extremes shortcut, we get $\frac{3x\cdot4}{5\cdot3x}$. We could multiply that out, but let's grab the copycat before it can hide!

• • • • • • • • • •

* But only if $x \neq 0$, $y \neq 0$, and $x \neq y$! (Because if any of that were true, then we'd get zero on the bottom of a fraction.)

Step 3. Rewriting it to show the copycat, we get $\frac{3x}{3x} \cdot \frac{4}{5} = \frac{4}{5}$. Well, it doesn't get much simpler than that, does it?

Answer: $\frac{4}{5}$

That's sort of interesting: No matter what you plug in for x,* that original expression will always equal $\frac{4}{5}$.

Doing the Math

Simplify these crazy-looking expressions. Assume no denominators equal zero. I'll do the first one for you.

1. $\frac{x}{x} - 1 + \dfrac{\frac{1}{x} + \frac{1}{y}}{\frac{1}{x} - \frac{1}{y}}$

<u>Working out the solution</u>: Yep, pretty crazy alright. But let's notice that the first part, $\frac{x}{x} - 1$, can be rewritten as $1 - 1 = 0$. Teachers do this to see if we're paying attention. Now we just have to deal with the scary part—no problem. Okay, from p. 66, we know that the bottom of the fraction, $\frac{1}{x} - \frac{1}{y}$, equals $\frac{y - x}{xy}$, so we'll keep that info in our back pocket. Now, let's simplify the numerator, and we get:

$\frac{1}{x} + \frac{1}{y} = \frac{y}{y} \cdot \frac{1}{x} + \frac{x}{x} \cdot \frac{1}{y} = \frac{y}{xy} + \frac{x}{xy} = \frac{y + x}{xy}$.

Our entire problem now looks like this: $0 + \dfrac{\frac{y + x}{xy}}{\frac{y - x}{xy}}$. Dropping

---

* Except zero! And that's because if $x = 0$ in the original expression, then the denominator would equal zero and that's simply not allowed in math. But you knew that.

the zero and using means and extremes, we get $\dfrac{xy(y+x)}{xy(y-x)}$.

I see a copycat: $\dfrac{xy}{xy} \cdot \dfrac{y+x}{y-x} = \dfrac{y+x}{y-x}$.

Answer: $\dfrac{y+x}{y-x}$

2. $\dfrac{\dfrac{3}{x} + \dfrac{x}{3}}{\dfrac{1}{3x}}$

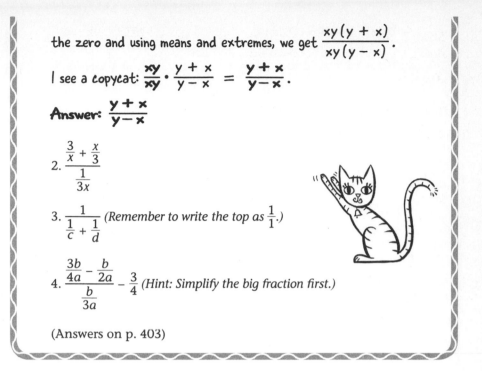

3. $\dfrac{1}{\dfrac{1}{c} + \dfrac{1}{d}}$ *(Remember to write the top as $\frac{1}{1}$.)*

4. $\dfrac{\dfrac{3b}{4a} - \dfrac{b}{2a}}{\dfrac{b}{3a}} - \dfrac{3}{4}$ *(Hint: Simplify the big fraction first.)*

(Answers on p. 403)

Well look at you, tackling these big scary problems! Nicely done. And remember, when things look so confusing that your brain shuts down (happens to the best of us), just pause for a moment, breathe, and take things one step at a time.

Reality Math

How to Outsmart Your Older Brother

Remember the dilemma from the beginning of the chapter? Your brother took $\frac{1}{3}$ of the first pizza, and he'll be back to take 2 pieces from the veggie pizza at any moment. How many slices should you cut the veggie pizza into so that when he takes two pieces of the veggie pizza, he'll only get his fair share, which is half a pizza total?

What's the best way to start? Label like a know-it-all!* Let's say we cut the veggie pizza into **v** slices. Then we could represent one *slice* by $\frac{1}{v}$ and two slices by $\frac{2}{v}$, right? And that's how much veggie pizza he'll get: $\frac{2}{v}$. For example, if you cut the veggie pizza into 6 pieces like the pepperoni, then $v = 6$, and he'd get $\frac{2}{6} = \frac{1}{3}$ of the veggie pizza. Or if you cut it into 200 pieces, he'd only end up with $\frac{2}{200} = \frac{1}{100}$ of the veggie pizza. (That might be going a bit too far.) Our goal is to have him end up with only half a pizza total—his fair share. Let's write this in math:

$$\frac{2}{6} + \frac{2}{v} = \frac{1}{2}$$

how much ↗
pepperoni he got

↑
how much
veggie
he'll get

↖ how much
total pizza
he'll get

Now we just need to figure out the value of **v** that makes this a true statement, and we'll have our answer! Our goal is to isolate *v* while keeping the two sides balanced. We haven't solved for variables much in this book yet, but you can follow along with your pre-algebra knowledge. (If you want, you can read Chapter 6 first and then come back.)

To start, we *could* combine the two fractions on the left, using a common denominator of **6v**, but it's actually faster to subtract $\frac{2}{6}$ from both sides, which also helps isolate *v*. So, $\frac{2}{6} + \frac{2}{v} = \frac{1}{2}$ becomes $\frac{2}{v} = \frac{1}{2} - \frac{2}{6}$. Using a common denominator of 6, the right side becomes $\frac{3}{6} - \frac{2}{6} = \frac{1}{6}$, so now our equation becomes:

$$\frac{2}{v} = \frac{1}{6}$$

You might already notice that if **v = 12**, then this is a true statement, after all, since $\frac{2}{12} = \frac{1}{6}$. But let's solve this the "real" way. Let's multiply both sides by 6*v* and watch the fractions disappear!

· · · · · · · · · ·
* See p. 9 to learn the benefits of being a know-it-all.

This will be our first step to isolating v.* Then we simplify:

$$\frac{2}{v} = \frac{1}{6} \quad \rightarrow \quad 6v \cdot \frac{2}{v} = 6v \cdot \frac{1}{6} \quad \rightarrow \quad \frac{12v}{v} = \frac{6v}{6} \quad \text{(find the copycats!)}$$

$$\rightarrow \quad \frac{12}{1} \cdot \frac{v}{v} = \frac{6}{6} \cdot \frac{v}{1} \quad \rightarrow \quad \frac{12}{1} = \frac{v}{1} \quad \rightarrow \quad v = 12$$

And voilà! We've found the v that makes our statement true. You should cut the veggie pizza into **12** slices. Then when he takes two slices, he'll get just $\frac{2}{12} = \frac{1}{6}$ of the veggie pizza, giving him a total of $\frac{2}{6} + \frac{1}{6} = \frac{3}{6}$: a half of a pizza, just like he should get! Ah, the crafty sister prevails. You leave out his 2 tiny $\frac{1}{12}$ slices on the kitchen table and return to your friends with the rest of the pizza, knowing that justice has been "served."

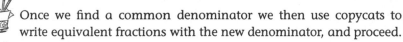

Takeaway Tips

Adding and subtracting fractions with variables works the same way as with numbers.

Once we find a common denominator we then use copycats to write equivalent fractions with the new denominator, and proceed.

Remember that you can rewrite things like x as $\frac{x}{1}$ or 5 as $\frac{5}{1}$ in order to more easily use copycats.

With a big complicated expression, most of the time your goal will be to rewrite it so it becomes a "manageable" complex fraction: a single fraction on top of a single fraction, ready for the means and extremes shortcut.

. . . .

* FYI, we multiplied both sides by $6v$—the product of the denominators—and that gave us the same result as if we'd cross-multiplied. Try it!

QUIZ: Are You Bold or Shy?

Everyone has fear and insecurities—everyone. But do you let yours get in the way of your life?

Take this quiz by expert psychologist Dr. Robyn Landow and contributor Anne Lowney, and see how you fare!

1. A school dance is coming up. The guy you've been crushing on all year seems pretty shy, so it might be totally up to you to do the asking! What do you do?

    a. You can't muster the courage to even talk to him. You're *way* too scared of rejection or embarrassment. What would he say? What would his friends say?

    b. Rather than flat out asking, which seems pretty scary, you strike up random conversations with him about the dance, hoping he'll get the hint and ask you.

    c. Hold your head up high, flash a smile, and ask him to the dance. With all that confidence, how could he say no?

2. You thought you understood the math lesson today, but then the teacher says something that seems wrong. You:

    a. Immediately raise your hand to point out the mistake. You want to make sure no one else gets confused by it.

    b. Think about it for a few minutes; yep, it still seems wrong. But you wait until after class to ask the teacher about it.

    c. Do nothing. Maybe someone else will ask the question.

3. You see a group of girls bullying your friend in the hallway. They're the girls we all hate: They tease, spread rumors and, if you try to confront them, make your life totally miserable. How do you handle the situation?

    a. Think a minute about what to do. You don't want to make enemies with those girls. You decide to walk by, call to your friend, and leave together. The leader of the crowd shoots you a dirty look, but you just keep walking.

    b. Approach the leader of their crowd directly and tell her what you think of her. Maybe you can shame her into thinking twice before bullying her next "victim."

c. You know it's wrong and you're totally leaving your friend alone, but you look the other way and pretend you didn't see anything. You are really afraid of being singled out—especially by *this* group.

4. Your school is putting on a production of *West Side Story*. Everyone is invited to sing in the chorus, but for a more significant part, a solo singing audition is required. That's right: you, stage, microphone, spotlight. What do you do?

    a. Go for the solo audition! You'll audition for Maria—the lead part.

    b. Nope—you shiver at the very *thought* of getting in front of a crowd.

    c. You consider the possibility, but you try for a smaller solo role.

5. You have a really good time in your foods class, so your teacher suggests you start up an after-school club. You:

    a. Wish you had the confidence to lead a new club. You're more likely to follow others than to be the leader. Maybe you'll get a friend to organize the whole thing.

    b. Think it's a great idea. You post banners, make sign-up sheets for the club, and make a formal request to use one of the common areas for meetings. This new visibility could even help your chances at running for student government!

    c. Invite a few close friends to start the club, but you're not exactly advertising. It's best to keep it small in case it doesn't work out.

6. Your parents are mad at you for not getting your chores done, but you've got a pile of homework that's stacked higher than the dishes in the sink. What do you do?

    a. State your case. No hesitation. You risk the chance of getting in more trouble, but you're the type of person who always defends herself. You know your priorities, and you'll do the chores tomorrow.

    b. Apologize and do the chores, wishing you had the guts to explain why the chores weren't getting done. Secretly it's an excuse to get out of studying for the history test. You can always blame a bad grade on chores later on.

    c. Complain about your homework, but, fearing their wrath, agree to do the chores—quickly—so you can get back to your homework. Both get compromised a little.

7. In terms of your social life, what is your greatest fear?

    a. Losing your most supportive friend.

    b. Any situation that singles you out to be judged by your peers, like public speaking.

    c. Losing the spotlight. You'd hate to fade into the background.

8. Your preferred method of communication is:

    a. Talking on the phone—one-on-one contact isn't always necessary.

    b. Talking in person—you're known for your charisma.

    c. Text messaging—it's a good way to avoid a potentially awkward conversation.

Scoring:

1. a = 3; b = 2; c = 1   4. a = 1; b = 3; c = 2   7. a = 2; b = 3; c = 1

2. a = 1; b = 2; c = 3   5. a = 3; b = 1; c = 2   8. a = 2; b = 1; c = 3

3. a = 2; b = 1; c = 3   6. a = 1; b = 3; c = 2

If you scored 8–12 points:

Good for you! You're the *queen* of charisma. You're not afraid to put yourself out there, and you're ambitious—which is one of the most important traits of a successful student and leader. Your confidence is infectious, and you make friends easily. Even guys respond positively to your self-assurance, and you probably have many of them as friends. However, be sure to remain respectful of authority: Your parents and teachers have a bit of life experience on you, after all. Also, as with all successful people, you will find that others will often be jealous. This doesn't mean you ever have to play small! But a word of caution: Don't be a show-off, and resist the urge to be a snitch or tattletale. Whenever you can, be generous; let someone else take the limelight when he or she deserves it, and go out of your way to point out others' accomplishments. They'll love you for it!

If you scored 13–18 points:

Although self-consciousness and fear of embarrassment can get in your way, you're still willing to try new things and move slightly out of your comfort zone from time to time. That's great! Now, take it a step further. Which of your deepest desires have you shied away from because of fear of failure or embarrassment? If you put yourself out there, what's the worst that could happen? Is the worst that could happen, like feeling embarrassed in front of your classmates, something you'd remember in 10 years? Probably not. How about the *best* thing that could happen, like you join the track team or become president of your class? Betcha the answer is yes—you would remember that! So what are you waiting for? Isn't it worth taking a risk? Be bold, girl! Here's a journal exercise: List the things you're good at and your best qualities. Generous? Kind? Intelligent? Then ask yourself if you're using those wonderful qualities to their full potential. You might be a nice and generous person, but if you're not courageous enough to give your teacher a compliment, she'll never know it! You might be smart, but if you are too shy to raise your hand in class, then you might stay confused about something that would otherwise have been easy. You have lots of talent and ideas, so let them shine!

If you scored 19–24 points:

Do you find yourself saying "sorry" when you haven't done anything wrong? And do you worry constantly about what other people think of you? It seems that the possibility of rejection or embarrassment is getting in the way of your dreams and relationships. You wish you could participate in more activities and meet more people, but your anxiety gets the best of you—yikes! Some extremely insecure people, like bullies, even overcompensate for their fear by going out of their way to make others feel bad about themselves. The truth is we ALL have fear; it's what we do with the fear that makes the difference. Try not to get bogged down by fear of other people's judgments. They're probably too busy secretly worrying what other people think of *them* to be paying attention to *you* anyway! Take it in baby steps: Walk into a room full of people with a big smile and say hello to the first person you see. Reach out to the new girl, join a club, or try out for a sports team. It might be hard at first, but the more you work on your confidence, the more natural it will become. Fake it till you make it! (Also, read the paragraph above this one. You'll find helpful stuff there, too.)

Chapter 6

Your Holiday Shopping List
Strategies for Solving for *X*

$\mathcal{H}$olidays are a great time. We get to shop for our family and friends, and it's always fun to pick out things we know they'll love. But it's so easy to spend more than we planned; those numbers sure do add up fast. Never mind the fact that the cutest shoes somehow ended up in our shopping cart. Now, how'd those get there?

But seriously, it can be hard to figure out how to stick to a budget, even if we don't splurge on ourselves. Later in this chapter, I'll show you how to make your very own custom algebra formula that will totally keep you on track for the holidays . . . as long as you stay away from those shoes.

We'll also look at some new strategies for solving for *x* that you probably didn't do in pre-algebra, and we'll use the skills we've been building from the last few chapters to do it. First, here's a mini review of the whole concept of solving for *x* to dust off the cobwebs in case it's been awhile since you've thought about this stuff.

Mini Review* of
Solving for x

When a math problem asks us to find the value of *x*, our goal is to isolate *x* by doing things to *both entire* sides of the equation—so that we keep the scales always balanced—until *x* is all by itself on one side, and a number is on the other side. That number is the answer!

When isolating *x*, it's helpful to think about *inverse operations*[†] . . . and gift wrapping.

Wrapping:
1. Put in box
2. Wrap with paper
3. Stick on sparkly bow

Unwrapping:
1. *Unstick* sparkly bow
2. *Unwrap* paper
3. *Take out* of box

Notice how the *last* thing we <u>did</u> to wrap the present is the *first* thing we <u>undo</u> to unwrap the present. Let's wrap up *x*! If we first multiply *x* times 2, we get **2x**. Then if we add 3, we get **2x + 3**. Then if we divide the whole thing by 5, we get $\frac{(2x + 3)}{5}$, right?

So to isolate *x*, we unwrap it by first doing the *inverse* of the last thing we did (dividing by 5), which is *multiplying* by 5, and we get **2x + 3**. Next, we do the inverse of adding 3, which is subtracting 3, and we get **2x**. Lastly, we do the inverse of the first thing we did (multiplying by 2), which is dividing by 2, and we get **x**, totally unwrapped!

---

* For a more in-depth review of the basics of solving for *x* and "keeping the scales balanced," check out Chapter 20 in *Math Doesn't Suck*.
† *Inverse operations* are operations that undo each other. Examples are addition & subtraction, multiplication & division, and opening & closing a box!

With equations, we use this same strategy to isolate *x*, but we must do those same operations to the other side of the equation, too, so that we can simultaneously unwrap *x* while *keeping the scales balanced*:

$$\frac{(2x + 3)}{5} = 3$$

$$(5)\frac{(2x + 3)}{5} = (5)3 \qquad \text{Multiply both sides by 5}$$

$$2x + 3 = 15$$

$$2x + 3 - 3 = 15 - 3 \qquad \text{Subtract 3 from both sides}$$

$$2x = 12$$

$$\frac{2x}{2} = \frac{12}{2} \qquad \text{Divide both sides by 2}$$

$$x = 6$$

And we end up with the answer, which we can plug back into the left-hand side of the original equation to check our work.

It's good to keep in mind that when we *unwrap x*, we're *undoing PEMDAS.** Notice that in this case, we first undid the only operation that *wasn't* in the parentheses, followed by addition/subtraction and finally multiplication/division. Yup, PEMDAS in reverse.

Note: This was a lightning-fast review of pre-algebra solving for *x*, inverse operations, and undoing PEMDAS. From this point on, I'm going to assume you're comfortable with solving these kinds of equations from pre-algebra. But if this chapter so far hasn't been totally review for you, then I highly recommend reading *Kiss My Math*, especially Chapter 12, which has a ton of great pre-algebra strategies that get used *constantly* in algebra.

Solving for One Variable in Terms of Another

Sometimes we need to solve for one variable "in terms of" another variable; for example, someone could ask us to solve for *x* in terms of *y*:

* See p. 400 in the Appendix for a review of the PEMDAS . . . and pandas!

$$2y + \frac{x}{3} = 5$$

And this can be weird, because when we isolate *x*, it's usually just numbers that we're isolating it from, right? So let's make it nice and use flowers 🏵 and smiley faces ☺, just to take the edge off.

$$2🏵 + \frac{☺}{3} = 5$$

Let's say we're told to solve for the smiley face in terms of the flower. That means we need to *isolate* the smiley face: We'll just <u>blindly pretend the flower is any other number</u> and collect it with the other numbers like we normally would. By subtracting 2🏵 from both sides, we get:

$$\frac{☺}{3} = 5 - 2🏵$$

And to isolate ☺, we multiply both sides by **3**:

$$3\left(\frac{☺}{3}\right) = 3(5 - 2🏵) \quad \rightarrow \quad ☺ = 15 - 6🏵$$

And we're done! So what does this *mean*? It means that if I tell you what number the flower is, you can instantly tell me what the smiley is. If I say, "The flower is equal to 0.5," then you could quickly plug in 0.5 where you see 🏵 and get ☺ = 15 − 6(0.5) = 15 − 3 = **12**. And that's what ☺ equals when 🏵 is 0.5.

We can also solve for the flower in terms of the smiley face. Now, we could start over with the original equation if we wanted, but the one we ended up with was so much nicer looking, and it's also a true statement. In fact, it's the same exact equation, just written differently! So, isolating the flower in the equation ☺ = 15 − 6🏵, we'll subtract 15 from both sides and then divide both sides by –6:

$$☺ - 15 = -6🏵 \quad \rightarrow \quad \frac{☺ - 15}{-6} = 🏵$$

Notice that if we multiply the fraction by $\frac{-1}{-1}$, we can make it look nicer:

$$\frac{-1(☺ - 15)}{-1(-6)} = \frac{-☺ + 15}{6} = \frac{15 - ☺}{6}$$

That means: 🏵 $= \dfrac{\mathbf{15 - ☺}}{\mathbf{6}}$, and we're done!

Turns out, we can solve for anything in terms of . . . anything.

Heck, let's go back to the equation, $\frac{☺}{3} = 5 - 2❀$, and solve for 5 in terms of the flowers and smiley faces. I mean, why not? It's a free country.

And we know what to do: To isolate 5, we'll need to add $2❀$ to both sides, right? And we get: $\frac{☺}{3} + 2❀ = 5$, in other words:

$$5 = \frac{☺}{3} + 2❀$$

I admit, it's kinda weird, but I wanted to show you that there are tons of ways to write the same true statement, and no matter what you're asked to solve for, you can do it. It's all a matter of perspective: Who's the special variable? And everything else is just "stuff" that we need to move out of the way so we can *isolate* our very special variable.

Doing the Math

a. <u>Solve for y</u> (in terms of x) and then evaluate at $x = 0, 2$.

b. <u>Solve for x</u> (in terms of y) and then evaluate at $y = 1, 9$. I'll do the first one for you.

1. $\dfrac{x + 2y}{3} = 6 + 3x$

<u>Working out the solution</u>: For part a, we shall remain calm and just focus on getting y by itself. So, let's multiply both sides by 3, and we'll get x + 2y = 18 + 9x, right? Now, subtracting x from both sides, we get 2y = 18 + 8x. Finally, we'll just divide both sides by 2: **y = 9 + 4x**. To finish part a, we'll plug x = 0 into this last equation, and we get y = **9**. When x = 2, then y = 9 + 4(2) = **17**. We're done with part a!

For part b, we don't have to start from the beginning; we can use our underlined equation above: **y = 9 + 4x**, which already looks so much simpler. To isolate x, let's subtract 9 from both sides and then divide by 4 to get:

$y = 9 + 4x \rightarrow y - 9 = 4x \rightarrow \dfrac{y-9}{4} = x$, which is

the same thing as $x = \dfrac{y-9}{4}$. To finish part b, let's plug

in the values of y that we were given: If y = 1, then

$x = \dfrac{1-9}{4} = \dfrac{-8}{4} = -2$. If y = 9, then $x = \dfrac{9-9}{4} = \dfrac{0}{4} = 0$.

Answer: a. y = 9 + 4x; 9, 17   b. $x = \dfrac{y-9}{4}$**; -2, 0**

2. $y + 4x = 2$

3. $\dfrac{x+1}{y} = 2$ *(Hint: Start by multiplying both sides by y.)*

4. $\dfrac{x+y}{13} = x$ *(Hint: Start by multiplying both sides by 13.)*

(Answers on p. 403)

It's All About Perspective

When you and your sister or brother (or friend) get into arguments, have you noticed how you can deal with it better once you've taken the time to see the situation from the other person's point of view? Solving for one variable in terms of another is like looking at an equation from one point of view, and then another. Solving for one variable in terms of another will come up constantly in word problems and graphing lines later in this book. It's a great skill to have—for math and for siblings!

Before we move on to some strategies for more challenging problems, I'd like to share a little holiday joy.

Reality Math

Your Holiday Shopping List: A Savvy Shopper Trick

Maybe your parents still buy presents for people "on your behalf," but soon enough, you'll be making your own shopping list, saving up your allowance, and buying presents for friends and family yourself. It's actually a really satisfying feeling!

Your list might be shorter or longer than this, but you get the basic idea. Let's say that after you've donated some toys to underprivileged kids, you've got $150 total left to spend on family and friends. Sounds easy enough, but problems start when you see the cutest dress for your BFF that costs $45, and then almost a third of your budget is gone. The next thing you know, you're making your sister a life-sized bird out of toothpicks. Ah, just what she wanted.

So . . . we're going to make a custom, designer formula to help you stay on budget and in control of how you distribute your money! The first thing you need to do is to decide the person you're going to spend the *least* amount of money on—like, say, your cousin. Sounds a little cruel, but c'mon, no one has to know . . . We'll call that amount "*x*." How much more than that

do you want to spend on each other person on the list? You might think to yourself, *Hmm, I'd probably want to spend about $10 **more** on my BFF than on my cousin.* Then the amount you'd spend on your BFF would be $x + 10$. And maybe Mom will be $x + 20$. See where I'm going with this? Just fill in each person's amount, in terms of *x*. Since we know the total equals $150, we can make a custom equation, and solve for *x!*

$$x + 20 + x + 20 + x + 15 + x + 10 + x = 150$$

Mom Dad sister BFF cousin

Now, let's combine like terms and solve!

$$\rightarrow 5x + 65 = 150$$

$$\rightarrow 5x = 85$$

$$\rightarrow x = \$17$$

And now, to find out how much you can afford to spend on each person, just plug in 17 wherever you see *x* on the list.

And voilà! A list that adds up to exactly your budget, $150, with easy guidelines for how much to spend on each person.

Of course, if you find something perfect for your cousin that costs $20, you can always subtract $3 from your BFF, or maybe you get an amazingly cute top for your mom that only costs $34. You'll end up adjusting the numbers in little ways as you shop, but it's so helpful to have this to gauge how you're doing. Then you'll know exactly *how* to adjust along the way. Learning how to stay on budget for anything is a great skill, and algebra puts *you* in the driver's seat.

Strategies for Solving for *x*

Sometimes teachers throw some pretty complicated-looking equations at us to solve. So I've compiled a list of helpful strategies for those little suckers. Let's tackle those problems that always seemed too scary or daunting before. You won't be easily intimidated, not as long as *I* have anything to say about it, girl. Bring 'em on!

More Solving for *x* in Terms of Another Variable

Let's say we're asked to solve for *x* in this equation.

$$3x - \frac{x}{2} + ax = 1 - \frac{3x}{4}$$

Hmm . . . fractions as coefficients are always sort of scary, and that *a* is a bit annoying, isn't it? Well, let's get rid of the fractions by multiplying both sides by 4.* Let's use the distributive property (say hi to everyone at the party!) and watch those negative signs:

$$4\left(3x - \frac{x}{2} + ax\right) = 4\left(1 - \frac{3x}{4}\right)$$

$$\rightarrow \quad 4(3x) - 4\left(\frac{x}{2}\right) + 4(ax) = 4(1) - 4\left(\frac{3x}{4}\right)$$

$$\rightarrow \quad 12x - 2x + 4ax = 4 - 3x$$

$$\rightarrow \quad 10x + 4ax = 4 - 3x$$

Okay . . . this looks better. The next step is to gather all the stuff with x to one side, right? I know, I know—we still have "Mr. a" hanging around; that's okay—just trust me. We'll add $3x$ to both sides, and we get:

$$\rightarrow \quad 13x + 4ax = 4$$

Now, all the stuff with x is together on one side, and it's as simplified as we can make it. Remember all that factoring we did in Chapter 3? We're going to now <u>factor out an x</u> by using reverse distribution: Yep, x is pulling out of the party.

$$13x + 4ax = 4$$

$$\rightarrow \quad x(13 + 4a) = 4$$

We've separated the x from "Mr. a," yay! And how do we finish isolating x? By dividing both sides by that stuff in the parentheses so that the x ends up by itself:

$$x(13 + 4a) = 4$$

$$\rightarrow \quad \frac{x(13 + 4a)}{(13 + 4a)} = \frac{4}{(13 + 4a)}$$

The (13+4a)'s cancel on the left fraction!

$$\rightarrow \quad x = \frac{4}{13 + 4a}$$

Done! And if we plugged that back into the original equation, it'd be messy, but we'd get a true statement.

Just to recap: Our strategy here was to collect the x's on one side. Then we used our new factoring skills (reverse distribution) from Chapter 3 to surgically remove the x from the a, so we could isolate and solve for x. We also practiced the strategies of multiplication and distribution to get rid of those fractions to begin with.

· · · · · · · · · ·

* Notice that 4 is the LCD of both fractions in the equation. That's why it works!

Speaking of fractions, here are some crazy ones like we saw in Chapter 5, and now we're going to solve *equations* with them.

Dealing with Big, Crazy Equations

At some point, you'll come across something like this:

$$\frac{\frac{3x}{5}}{\frac{x}{4} + \frac{x}{2}} = \frac{x}{5} + 3$$

Totally crazy, right? Well, the key behind solving these kinds of wacky fraction problems is to break them into bite-sized pieces. We deal with the numerator and denominator *separately*. In fact, hmm, we've already dealt with this very fraction (see p. 67)! So using the same strategy, this equation simplifies into:

$$\frac{4}{5} = \frac{x}{5} + 3$$

Ah, looking much better. I feel like getting rid of the fraction element completely, don't you? Let's multiply both sides by 5:

$$5\left(\frac{4}{5}\right) = 5\left(\frac{x}{5} + 3\right)$$

$$\rightarrow 4 = x + 15$$

And now we just subtract 15 from both sides and get **$x = -11$**.

Answer: $x = -11$

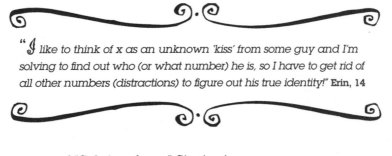

"*I* like to think of x as an unknown 'kiss' from some guy and I'm solving to find out who (or what number) he is, so I have to get rid of all other numbers (distractions) to figure out his true identity!" Erin, 14

Summary of "Solving for *x*" Strategies

When solving for *x* (or any problem in life!), it's not always so clear what to do next, is it? Well, you're growing up, and so is the math: There will often be many ways to tackle these problems. Below is a summary of my favorite ways to handle some of the sticky issues that come up in equations, many of which we've seen in the last few pages. But every problem is different, so use these tools when it makes sense to *you*!* And remember, above all, our goal is to isolate *x* while keeping the scales balanced.

Complaint	Possible Strategy
The coefficients are fractions. I don't like it when the coefficients are fractions.	Use the distributive property to multiply both sides by the LCM of the denominators, so the fractions go away.
I don't like all these parentheses!	Use the distributive property to multiply stuff out.
What's with all the negative signs?	Multiply both sides by –1, and use the distributive property to do it right!
Um, one (or more) of the *x*'s is stuck to another *variable*.	Gather all the terms with *x* in them, and <u>factor out the *x*</u>. (See pp. 83–84.)
Hey, there's an *x* on the bottom of a fraction. Help?	Multiply both sides by the denominator. Warning: You *must* check your answer, and denominators can't equal 0!
The coefficients have decimals in them, and it's kind of freaking me out.	Multiply both sides by 10, or 100, or 1000, etc. (We'll practice this in Chapter 16.)
There is a certifiably crazy fraction on one side of this equation.	Before even starting to solve, simplify the numerator and denominator *separately.* Then create a simple fraction from it, and *then* move on to the solving part!

* Also check out the "Solving for *x* Workshop" on pp. 183–87 in *Kiss My Math*.

Watch Out!

Every time that I say "multiply both sides by" something, remember to use parentheses and the distributive property. It's important that we multiply both <u>entire</u> sides by the same thing to keep the scales balanced. Otherwise, we'd end up with an untrue statement, and when we were finally left with "x = something," it would be untrue, too!

Let's lift some heavy weights and use the above strategies to solve these crazy-looking equations. Some will have other variables in them, and some won't. Look, if you didn't have to struggle with these at least a little, you wouldn't be normal. It's how we *deal* with fear that separates the strong young women from the little girls. Stick with me, sister, and you won't be easily scared off!

Doing the Math

Solve for **x**. For a challenge, try them without my hints! I'll do the first one for you.

1. $\dfrac{ab}{\frac{1}{2}a} + bx = 0.7x$

<u>Working out the solution</u>: Even without knowing how this is going to turn out, let's bring all the stuff with x together, so we'll subtract bx from both sides, and get $\dfrac{ab}{\frac{1}{2}a} = 0.7x - bx$.

On p. 54, we used complex fractions to learn that the left side equals 2b, so now our equation looks like this: 2b = 0.7x - bx. We can't keep that decimal around, so multiplying both sides by 10 gives us 10(2b) = 10(0.7x - bx) → 20b = 7x - 10bx. Looking better! To isolate x, let's factor out x with reverse distribution and write our equation like this: $20b = 7x - 10bx$ → $20b = x(7 - 10b)$. To finish isolating x, we'll divide both sides by $(7 - 10b)$ and get

$\dfrac{20b}{7 - 10b} = x$. And that's our answer!

Answer: $x = \dfrac{20b}{7 - 10b}$

2. $ax + bx = c$ *(Hint: Factor out x first!)*

3. $x(y + z + 3) = 7(x + 2)$ *(Hint: First distribute the x and 7, then collect the stuff with x's to one side, and then factor out the x.)*

4. $\dfrac{d(c - 2)}{(2 - c)} + dx = 5$ *(Hint: First, simplify just the fraction. See pp. 48–49.)*

5. $\dfrac{\frac{x}{2} + \frac{x}{6}}{\frac{2}{3}} = 2x - 1$ *(Hint: First, simplify just the complex fraction like we did on p. 67.)*

(Answers on p. 403)

Takeaway Tips

No matter how complicated things look, your goal is to isolate the variable you're solving for. Collect those on one side and everything else on the other side.

If *x* is stuck to other variables with multiplication, then after you gather the *x*'s to one side, factor out *x* with reverse distribution, like we learned in Chapter 3.

Every problem is different! And *practice* is the best way to get good at solving for *x*.

Danica's Diary
50 THINGS THIS WEEK

In the tenth grade, I used to worry about getting all my work done as well as everything else I had scheduled for the week. I was acting on the TV show *The Wonder Years*, so I had lines to memorize on top of all my school assignments and other commitments. But I wasn't alone; I had friends on the school paper and sports teams who felt just as overwhelmed as I did—maybe more so!

I was lucky to have a really great tenth-grade history teacher who seemed to know what we were all going through. Her name was Mrs. Hof, and she would often begin class by telling us a story about her life, usually having to do with issues of stress and time management. She was in the process of buying a house, selling her old house, teaching, grading papers, and driving her daughter to school and gymnastics, among other things.

She told us that when she felt overwhelmed with too many things to do, she might think, "I need to get 50 things done this week, and I don't know how to do 50 things in one week." But then a moment later, she'd realize, ". . . but I can do these 10 things today. That much I know I can do. And you know what? I don't have to worry *at all* about the other 40 things; I can really just focus on these 10 things for today. That's it, 10 things."

I still think about that—all the time.

When I got my first book deal, for *Math Doesn't Suck*, I was really excited. I was going to be a real live book author! But then I was like, "Wait, a 300-page book? Even if I have six months, I have no idea how to write a 300-page book!"

And then I remembered Mrs. Hof's method of breaking it down into smaller, bite-sized pieces of work, and I thought, "Hmm. . . let's see. If it takes me about 3 weeks to write some sort of outline for it and brainstorm ideas, then I'll have 21 weeks left to write 300 pages. That works out to just over 14 pages a week. But let's say I only work 5 days per week. Then I'd need to write 3 pages each day to be safe."

Just 3 pages? Well! That sure sounded better than 300 pages. Now some days I wrote 5 or 6 or even 10 pages if I was really on a roll. And some days I got busy with other things and didn't get to write at all, or I spent the whole day *rewriting* pages so I didn't add any new pages to the book. And whenever I felt panic creeping up, I'd think about Mrs. Hof and her strategy. And it worked!

You can apply this strategy to just about anything, whether it's a huge class history project or a long biology report, or even a multistep, scary looking math problem.* Don't let the *size* of a task scare you off. That's how a lot of people end up procrastinating and then cramming: They feel too scared to approach such a big endeavor; they'd rather hide from it, and then they end up with even less time, more stress, and a guilty conscience for procrastinating.

The solution? Break it down into *smaller pieces*, and for now, focus *only* on the small piece in front of you. Don't worry about the entire project, just the piece in front of you, and do that one piece *now*. It really works!

As one of my ninth-grade teachers, Mr. Coombs, was fond of quoting, "The journey of a thousand miles begins with a single step."

I'm currently writing my third book, which is an enormous project. It's easy to feel overwhelmed when there's so much to do, but I've learned that breaking it down into bite-sized pieces and focusing on those one at a time makes all the difference. And if you're reading this...then I guess it worked!

. . . .

* To see what I mean about breaking down a *math problem* into bite-sized pieces, just see the problem on p. 68.

Chapter 7

How Picky Are You?
Solving Equations with Absolute Value

$\mathcal{L}$et's say you felt like seeing a movie, but when you got to the theater, it was sold out. Are you the type to get bummed out and go home, or would you be willing to check out the other movies and see if any of them might satisfy  your movie craving? In other words, do you tend to be satisfied by *more* than one option, or do you have to have the *one* thing you had your mind set on?

It's important to have standards, but people who are super picky don't tend to be as happy in life. They decide they can be satisfied by only one option, and then nothing else *can* satisfy them. I suppose some people are just pickier than others.

Well guess what? Some *equations* are pickier than others, too . . .

So far, we've been solving equations that are only satisfied by one value of x. In the equation $2x + 1 = 3$, plugging in $x = 1$ creates a <u>true statement</u>, and that is the *only* value that will! If you plugged in any other value besides $x = 1$, you'd get a false statement.

Now we're going to look at one type of equation that isn't quite so picky: equations with **absolute value** bars surrounding the variable. After all, most of these equations are satisfied by *two* values. They don't even care which one; either is fine with them.

The **absolute value** of a number or expression is its <u>distance</u> to zero on the number line, so it will never be negative; for example, $|3| = 3$, $|-6| = 6$, $|0| = 0$, and $|3 - 8| = 5$.

Absolute Value: A Luxury Spa

I like to think of the inside of the absolute value bars as a nice, relaxing place like a luxury spa, because when a number comes out, it's so happy and positive!*

Similarly, *equations* with absolute value bars surrounding the variable are much more easygoing than normal equations. I mean, spas usually put people in a pretty good mood, so you can see how these equations might be a little more relaxed about the movies they want to see, or the values that will satisfy them. (This really does make perfect sense.)

Before we solve these kinds of equations, first let's see what it means to have absolute value bars around a variable. Because $|-2| = 2$ and $|2| = 2$, then for the equation $|x| = 2$, we can see that x could have either of *two* values (2 or –2), and this equation would indeed be satisfied either way. So the solution is **$x = -2$ or 2**. Make sense?

Taking it one step further, if we are given $|x + 1| = 2$, then what values of x would satisfy this? As it turns out, x could be either 1 or –3. It's easy to see how $x = 1$ satisfies the equation, since $|\mathbf{1} + 1| = \mathbf{2}$. Or instead, if $x = -3$, we get $|\mathbf{-3} + 1| = |-2| = \mathbf{2}$. Well how about that! Our equation is a true statement for two values: **$x = -3$ or 1**.

You see, we found values of x that made the *inside* of those bars equal 2 or –2. And then the entire equation was satisfied, because both $|2| = 2$ and $|-2| = 2$. See what I mean? Let me show you how this works.

Solving Equations with Absolute Values

When we have an equation that looks like this:

$$\left| \text{ something with a variable in it } \right| = \text{ number}$$

. . . .

* To read more on absolute values (and spas), see Chapter 4 in *Kiss My Math*.

... then as long as the "number" is *positive,* there will usually be *two* values that satisfy the equation! And here's how we find those values: We look at just what's <u>inside</u> the bars—the inside "guts"—and then create two normal equations to solve: one equation where the <u>inside guts</u> equals the number, and one where the <u>inside guts</u> equals the *opposite* of the number. You see, both types of inside guts, negative *and* positive, will satisfy the original absolute value equation. Let's do an example with this strategy:

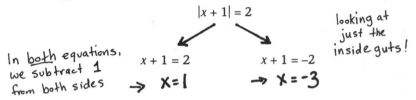

$$|x + 1| = 2$$

looking at just the inside guts!

In <u>both</u> equations, we subtract 1 from both sides

$$x + 1 = 2 \rightarrow x = 1$$

$$x + 1 = -2 \rightarrow x = -3$$

Creating these two equations is the trick; solving them is the easy part. We already checked these on p. 92, so we're done!

Answer: $x = 1, -3$

QUICK (REMINDER) NOTE   Absolute value bars are a grouping symbol, so they have the same priority as parentheses when it comes to the PEMDAS order of operations. (See p. 400 for PEMDAS.)

With something a little more complicated like this, $3|x + 1| = 6$, our first step is to get the <u>absolute value expression</u> by itself, by dividing both sides by 3. Doing this (and remembering that absolute value bars act much like parentheses*), we get:

$$3|x + 1| = 6 \quad \rightarrow \quad \frac{3|x + 1|}{3} = \frac{6}{3} \quad \rightarrow \quad |x + 1| = 2$$

And we'd proceed as normal! Here's the method, step by step:

. . . .

* But never distribute across absolute value bars. The distributive property (pp. 30–31) only works across parentheses ( ) and brackets [ ].

Step By Step

Solving absolute value equations:

Step 1. First, verify that the absolute value bars contain a variable. If they do, then isolate the bars (without disturbing what's *inside* them) until the equation looks like this:

|something with a variable in it| = a positive number

(If the number on the right *isn't* positive, see the QUICK NOTE below.)

Step 2. Create two equations by looking at what's inside the bars: For equation #1, set the <u>inside guts</u> equal to the number on the right. For equation #2, set the <u>inside guts</u> equal to the *negative* of the number on the right.

Step 3. Solve each equation. Now you've got the two values of *x* that satisfy the original (absolute value) equation. Check your work by plugging the *x* values back into the original equation. Done!

QUICK NOTE When you put the equation into the correct form: |something with a variable in it| = number . . . if the "number" equals zero, then you'll end up with just *one* answer, not two. Because −0 = 0, the "two" equations you split off and solve are actually the same equation! On the other hand, once the equation is in the above form, if the "number" is negative, then there are NO solutions to the problem. After all, absolute values are *never* negative.

And... Action! Step By Step In Action

Find the values of x that satisfy the equation $2|x - 7| + 3 = 13$.

Step 1. Let's isolate the absolute value bars. First, we'll subtract 3 from both sides and get $2|x - 7| = 10$. Then, we'll divide both sides by 2 and get $|x - 7| = 5$. Good! We've isolated the bars.

Steps 2 and 3. Now we'll create two equations from the "inside guts":

$$|x - 7| = 5$$

 just the inside guts!

$$x - 7 = 5 \qquad x - 7 = -5$$

For equation #1, we just add 7 to both sides, and we get **$x = 12$**. For equation #2, we also add 7 to both sides and get **$x = 2$**. Checking our work, we'll first plug **12** into the left side of the *original* equation and hope to get 13. So, $2|12 - 7| + 3 = 2(5) + 3 = 13$. It worked! Plugging in **2**, we get $2|2 - 7| + 3 = 2|-5| + 3 = 2(5) + 3 = 13$. Worked again!

Answer: $x = 2, 12$

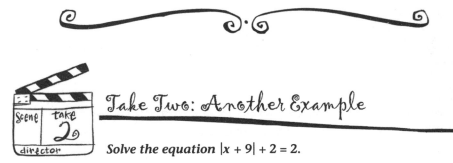

"*J*ust like a smile, mathematics is a powerful tool that transcends cultures. Two people of any age, race, or gender can share in the language of mathematics—even if they do not share in the same verbal language. After all, 2 + 2 is the same on every continent!"

Courtney, 18

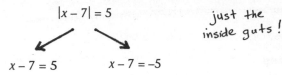

Take Two: Another Example

Solve the equation $|x + 9| + 2 = 2$.

Step 1. Let's subtract 2 from both sides and get $|x + 9| = 0$.

Steps 2 and 3. Now that the bars are isolated, we are ready to create our two equations. For the first equation, we'll set the insides equal to 0, so we get $x + 9 = 0$. Subtracting 9 from both sides, we get our answer: **$x = -9$**. For our second equation, we set the insides equal to –0. But wait, $-0 = 0$, so our second equation would be the same. That means there's only one solution to this equation. (See also the QUICK NOTE on p. 94.) Plugging our answer into the left side of the equation, we're hoping to get 2: $|-9 + 9| + 2 = 0 + 2 = 2$. Great!

Answer: $x = -9$

Watch Out!

Do Not Disturb!

Don't ever EVER "make" something positive when you have a variable inside absolute value signs. *Ever.* Here's what I mean: If you see $|-x + 3| = 2$, don't even THINK about taking off the negative sign. You must simply take the inside guts—*whatever* they look like—and use them to create your two equations. Would you ever dream of disturbing someone getting a massage? I didn't think so.

QUICK NOTE   It's NOT necessarily true that $|-x| = x$. In fact, it's just as likely that $|-x| = -x$. Remember, x *itself* might be negative, which would mean that $-x$ would be a *positive* number. It's another good reason never to disturb the inside of the absolute value bars when there's a variable inside; we just don't know what we're dealing with . . .

Take Three: Yet Another Example

Let's find the values of x that make this equation true:

$$|-x + 2| = 3$$

We'd never *dream* of taking off that leading negative sign, right? (Do not disturb!)

Steps 1 and 2. Yep, it's in the right form, with the absolute value expression isolated. Let's create our two equations:

$$|-x + 2| = 3$$

just the inside guts!

$$-x + 2 = 3 \qquad -x + 2 = -3$$

Step 3. For equation #1, let's subtract 2 from both sides, and we get $-x = 1$, which means $\boldsymbol{x = -1}$. For equation #2, we'll also subtract 2 from both sides, and we get $-x = -5$, so $\boldsymbol{x = 5}$. Let's check 'em: Plugging $x = -1$ into the <u>left side</u> of the original equation, we hope to get 3, right? $|-(\boldsymbol{-1}) + 2| = |1 + 2| = |3| = 3$. Great! Plugging in $x = 5$, we get $|-(\boldsymbol{5}) + 2| = |-3|$ $= 3$. Yep, we've found both values that satisfy the equation!

Answer: $x = -1, 5$

QUICK NOTE   It's really easy to graph the solutions of absolute value equations on the number line. Most of the time, it'll just be two dots.

Doing the Math

Find both values that satisfy these equations and graph their solutions on the number line. I'll do the first one for you.

1. $\dfrac{|3 - y|}{4} = 1$

<u>Working out the solution:</u> First, let's isolate the bars by multiplying both sides by 4:

$$\frac{4|3 - y|}{4} = 4(1) \quad \rightarrow \quad |3 - y| = 4$$

Using the inside guts, $3 - y$, equation #1 will be $3 - y = 4$. To isolate y, let's subtract 3 from both sides to get $-y = 1$. Then we multiply both sides by -1 to get $\boldsymbol{y = -1}$. Using the inside guts again, equation #2 will be $3 - y = -4$. Subtracting 3 from both sides and multiplying both sides by -1, we get $-y = -7 \rightarrow \boldsymbol{y = 7}$.

Plugging either y = −1 or 7 into the original equation will give a true statement (try it!), so we have found both values of y that will satisfy the equation.

Answer: y = −1, 7

-10 -9 -8 -7 -6 -5 -4 -3 -2 -1 | 0 1 2 3 4 5 6 **7** 8 9 10

$y = -1, 7$

2. $|a - 3| = 6$

3. $|-a - 3| = 6$

4. $\dfrac{|b - 2|}{2} = 1$

5. $3|h + 5| - 1 = 8$

(Answers on p. 403)

ℑakeaway ℑips

Equations with absolute values usually have *two* solutions. In other words, the equation is satisfied by two different values of *x* (or whatever variable is being used).

To solve these equations, first isolate the absolute value bars, and then use the "inside guts": Set them equal to the number on the right and also to the *negative* of the number on the right, and you'll get two non-absolute value equations to solve.

Remember that |−*x*| does not necessarily equal *x*, because *x itself* could be a negative number. In that case, −*x* would be a positive number. So never remove negative signs from inside the bars when a variable is involved.

Chapter 8

Control Freaks, Unite!

Solving Inequalities with Absolute Value

Do you cram a gazillion things into each day, texting in between each one? Ever spend all day at the beach—on the phone? Ever find yourself wishing you could control everything and everyone around you, all the time? This is when it's time to stop, take a breath, and chill—maybe do some yoga, perhaps even get a massage.* I mean, being a fabulous multitasking gal in control of her own destiny can be great, but life is so much more satisfying when we put down our electronics from time to time and just relax.

There was this girl Connie who used to struggle with just that, and her controlling ways will help you remember how to solve inequalities with absolute values. I'll tell you all about her in a moment. For now, let's review inequalities, shall we?

. . . .

* My mom and I made a yoga/meditation DVD called *Daily Dose of Dharma*. And you know what else? Any time of day, you can give *yourself* a little foot massage. It's great!

Reviewing Inequalities

Remember inequalities?

$$x > -3$$

Most people agree that the inequality symbols $<$, $>$, $\leq$, and $\geq$ look like hungry alligator mouths, pointing to the bigger value, so $x > -3$ means that x can have all values bigger than -3. And to graph inequalities, we just draw an open or closed dot* and a ray extending in one direction. Solving inequalities is almost exactly the same as solving equations; we do things to both sides until there's a single variable on one side and a number on the other. The only difference is what I like to call the Mirror rule:

> **The Mirror Rule for Inequality Symbols: When we _multiply_ or _divide_ both sides of an inequality by a negative number, we must REVERSE the direction of the inequality symbol.**

For example, to solve $-x + 2 > 7$, first we'd subtract 2 from both sides to get $-x > 5$, and then we'd multiply both sides by -1 and also switch the direction of the symbol to get our answer: $\boldsymbol{x < -5}$.

To check our work, we plug in <u>any</u> value less than -5, like how about $x = -10$, into the original equation and hope to get a true statement:
$$-x + 2 > 7 \;\rightarrow\; -(\boldsymbol{-10}) + 2 \overset{?}{>} 7 \;\rightarrow\; 10 + 2 \overset{?}{>} 7 \;\rightarrow\; 12 > 7. \text{ Yep!}$$

• • • • • • • • • •

* Open dots are used for $<$ and $>$, and closed dots are used for $\leq$ and $\geq$.

This was a super-fast review of solving inequalities; for more, check out Chapter 14 in *Kiss My Math*. Before we get back to Connie, let's also review **solution sets**.

Ring Ring

What's It Called?

The **solution set** is the collection of all values that satisfy a given condition. For example:

The solution set of $x + 1 = 3$ is **x = 2**:

$$x = 2$$

The solution set of $|x| = 2$ is **x = –2, 2**:

$$x = -2, 2$$

The solution set of $x + 1 < 3$ is **x < 2**:

$$x < 2$$

The solution set of $|x| < 2$ is **–2 < x < 2**:

$$-2 < x < 2$$

(We'll see this last one on p. 106.)

Combining Inequalities . . .
and Scheduling a Massage

So there was this girl Connie who was a total control freak. But even she knew that she had to relax, so she called up a spa one day and was like, "I'm totally busy but I need a massage; my 1-hour massage can't start before 3 P.M., but also, my massage would have to start no later than 5 P.M., because I have a business dinner to attend."

Her appointment has two conditions on it. Let's call the <u>start time</u> t. She has said that $t \geq 3$ and $t \leq 5$, and *both* must be true for the massage to fit into her schedule, right? So we can graph the start times that work for Connie like this:

Connie:

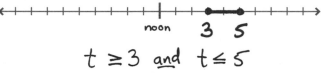

$$t \geq 3 \text{ and } t \leq 5$$

There was this other girl Daisy, who wasn't so busy. She called up the spa and said, "Hey, so my massage could be 2 o'clock or earlier, or it could start at 4 o'clock or later. Whatever works best for you guys." Her start time has two options: $t \leq 2$ or $t \geq 4$, so we can graph all the times that work for Daisy like this:

Daisy:

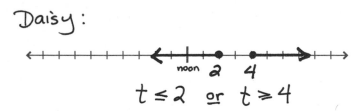

$$t \leq 2 \text{ or } t \geq 4$$

We've just seen two ways we can combine inequalities. Connie's combo of requirements is called a **conjunction**, and Daisy's is called a **disjunction.***

. . . .

* Yeah, I tried to think of a common girl's name that starts with "Dis" but no such luck.

What's It Called?

A **conjunction** is a statement of two conditions connected by the word *and*, like this:

$$x > 3 \quad \text{and} \quad x < 5$$

. . . this can also be written in this more compact form:

$$3 < x < 5$$

For a conjunction to be satisfied, **both** its conditions must be satisfied. Conjunctions are somewhat controlling like that, aren't they? (They're also very busy so they need to be efficient, which explains why they came up with a more compact form.)

A **disjunction** is a statement of two conditions connected by the word *or*, like this:

$$x < 2 \quad \text{or} \quad x > 4$$

(There is no compact form for disjunctions.) If either of its two conditions can be satisfied, then the disjunction will be satisfied. Disjunctions are very easygoing and nonchalant like that, almost to the point of being distracted.

Unfortunately, Daisy just isn't as organized as Connie, and Daisy ended up forgetting all about her massage appointment. Connie got her massage, but she was so tense about making it to dinner on time that she didn't even enjoy it!

In math, you're either a conjunction or a disjunction. In life, let's strike a balance, okay?

Solving for *x*, and Synchronized Swimmers

When we solve a disjunction like this:

$$2x + 1 \leq 5 \quad \text{or} \quad 3x \geq 12$$

. . . we just solve each little inequality separately and then graph them on a number line. These are easy to solve; we get **$x \leq 2$ or $x \geq 4$**. (See previous page for its graph, using *t* instead of *x*.)

When we solve a *conjunction*, like $1 \geq -4 - n > -7$, we have two choices about how to solve it. We can either break it up into two inequalities, or we can solve the whole thing at once. In this latter case, instead of doing the same thing to both sides of an inequality to solve it, we do the same thing to all *three* sections at once. This always reminds me of three little synchronized swimmers, with their perfectly synchronized backstrokes and smiling faces. (How do they smile while they're holding their breath?)

$$1 \geq -4 - n > -7$$

To isolate n, we'll add 4 to all three of our little swimmers, in unison:

$$1 + \mathbf{4} \geq -4 + \mathbf{4} - n > -7 + \mathbf{4}$$

$$\rightarrow \quad 5 \geq -n > -3$$

Now, let's multiply everyone by -1, and the Mirror rule says to flip the inequality symbols.

$$-5 \leq n < 3$$

Because n has been isolated, we're ready to graph n's values—our solution set.

Answer:

-6 -5 -4 -3 -2 -1 **0** 1 2 3 4 5 6

$$-5 \leq n < 3$$

QUICK NOTE   Some **conjunctions** will have two requirements that don't match up, like:

$$p > 2 \text{ and } p < 1$$

Let's face it: If a number is greater than 2, it can't *also* be less than 1. Since no value of p could ever fulfill both requirements, there is <u>no solution</u>. There's nothing to graph! On the other hand, some **disjunctions** may have two requirements that overlap, like:

$$p > 1 \text{ or } p < 2$$

In this case, the solution set is all the real numbers, because every single real number satisfies at least one of these conditions. Try it! One million satisfies the first one, since one million > 1, and −2000 satisfies the second one, since −2000 < 2. (Some numbers, like 1.5, actually satisfy <u>both</u>. Show-offs.) The graph of values satisfying this disjunction is the entire number line:

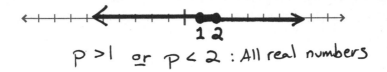

$p > 1$ <u>or</u> $p < 2$: All real numbers

Doing the Math

Solve each sentence, say if it's a conjunction or a disjunction, and graph the solution set. I'll do the first one for you.

1. $-7 \le -1 + z < 0$

<u>Working out the solution</u>: Time to do some synchronized swimming! To isolate z, we just add 1 to everyone, and we get **$-6 \le z < 1$.** Now we're ready to graph it, remembering we'll need one closed dot and one open dot.

Answer: It's a conjunction.

$$-6 \le z < 1$$

2. $n \le 3$ or $n > 6$

3. $n \le 3$ and $n > 6$

4. $-5 < 2y + 3 \le 1$

5. $-2 \le w - 7 < 1$

(Answers on p. 403)

Inequalities with Absolute Value

We saw absolute value equations in the previous chapter. They weren't nearly as picky as normal equations, because they could be satisfied by two values, remember? Now what about absolute value *inequalities,* like $|z - 3| \leq 1$?

First, let's consider something much simpler. Remember that absolute value represents the inside's *distance to zero* on the number line. So, for example, $|x| < 2$ represents all values whose distance from zero is less than 2.

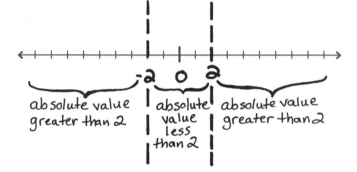

We can see from the graph that $|x| < 2$ means that x must be greater than –2 and also less than 2. In other words, $|x| < 2$ is the same as the conjunction $-2 < x < 2$.

Similarly, $|x| > 2$ means that x must be either less than –2, or it could be greater than 2. In other words, $|x| > 2$ is the same as the disjunction "$x < -2$ or $x > 2$."

When you're solving absolute value inequalities, the solutions will almost always be conjunctions or disjunctions.* But don't get bogged down by the words *conjunction* and *disjunction*. The most important thing to know about your solution is <u>what its graph looks like</u>: Is it the controlling kind of inequality satisfied only by a line segment connected by two boundary points? Or is the solution more easygoing, made up of two separate rays, going to infinity in both directions?

· · · · · · · · · ·

* For the exceptions to this, see the mutant cases on pp. 111–12.

Our strategy for absolute value inequalities will be this: First, we'll solve them as if they had equals signs, and this will give us our two boundary points (the spots on the number line where the dots go). Then, we'll figure out if our solution is the segment *between* the two points or the stuff on the *outside* of them. Here's the step by step!

Step By Step

Solving absolute value inequalities:

Step 1. Simplify the inequality by isolating the absolute value expression until you get "|something with a variable| < number" (or with >, ≤, or ≥). If the "number" is zero or negative, <u>stop these steps</u> and see the What's the Deal on pp. 111–12.*

Step 2. Assuming the "number" above is indeed positive, then it's time to find the two boundary points. Replace the inequality symbol with an equals sign, and solve using the step-by-step method from p. 94.

Step 3. Graph these two boundary points on a number line. If the original inequality symbol was < or >, then both dots will be open. If it was ≤ or ≥, then both dots will be closed. These two boundary points divide the number line into three regions.

Step 4. Next, pick an easy value *in between* your two boundary points (don't pick a boundary point) and plug it into the inequality you got at the end of Step 1. If it's satisfied by the value, then draw a line between the boundary points; the answer is a conjunction. If not, then double-check to make sure that points in the outer regions satisfy the inequality, and draw two rays extending in either direction; the answer is a disjunction.

Step 5. Write the answer above or underneath the number line. Done!

. . . .

* Advanced footnote: In this book we won't consider expressions with a compound use of absolute value, such as |–|x| + 2| = 0. So, assume that "something with a variable" won't ever include more absolute value bars inside it.

Solve this inequality and graph the solution set: $|z - 2| \leq 3$.

Step 1. It's already in the correct form with a positive number on the right. Great!

Step 2. Let's find the boundary points. We'll replace the $\leq$ with $=$ and solve the equation $|z - 2| = 3$, using the method from the previous chapter.

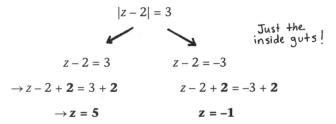

$$|z - 2| = 3$$

Just the inside guts!

$z - 2 = 3$	$z - 2 = -3$
$\rightarrow z - 2 + \mathbf{2} = 3 + \mathbf{2}$	$z - 2 + \mathbf{2} = -3 + \mathbf{2}$
$\rightarrow z = 5$	$z = -1$

Step 3. Our boundary points are **z = –1, 5**. They will both be <u>closed</u> dots because our inequality was $\leq$.

$$\overset{\bullet}{\underset{-1}{}} \quad \overset{\bullet}{\underset{5}{}}$$

Step 4. Now we'll pick an easy value between the boundary points to test. Let's use $z = 0$, and see if it satisfies the original inequality: $|z - 2| \leq 3 \rightarrow |0 - 2| \overset{?}{\leq} 3 \rightarrow |-2| \overset{?}{\leq} 3 \rightarrow 2 \leq 3$. Yep! This means the entire region between the boundary points also satisfies the inequality. (We can see how values outside the boundary points don't work, like $z = 10$: $|10 - 2| \overset{?}{\leq} 3 \rightarrow 8 \overset{?}{\leq} 3$ Nope!)

Step 5. We've found that the region between –1 and 5 satisfies the inequality, so the solution is $\mathbf{-1 \leq z \leq 5}$.

Answer:

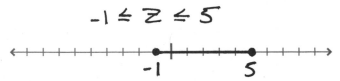

$$-1 \leq z \leq 5$$

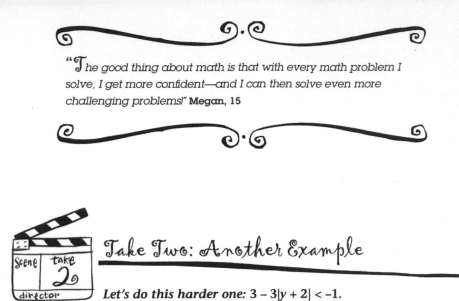

"The good thing about math is that with every math problem I solve, I get more confident—and I can then solve even more challenging problems!" **Megan, 15**

Take Two: Another Example

Let's do this harder one: $3 - 3|y + 2| < -1$.

Step 1. It might seem at first that this has no solution because of the –1, but don't judge it yet! To isolate the absolute value bars, we'll first subtract 3 from both sides and we get $-3|y + 2| < -4$. Then let's divide both sides by –3. The Mirror rule says we then also have to reverse the direction of the inequality. Let's do it:

$$-3|y + 2| < -4$$

$$\rightarrow \quad \frac{-3|y + 2|}{-3} > \frac{-4}{-3}$$

$$\rightarrow \quad |y + 2| > \frac{4}{3}$$

We've isolated the absolute value bars, and our number, $\frac{4}{3}$, is indeed positive.

Step 2. Let's find the boundary points by replacing > with =, and then creating our two equations, using the inside guts:

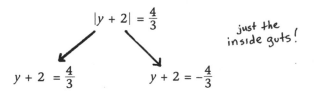

$$|y + 2| = \frac{4}{3}$$

$$y + 2 = \frac{4}{3} \qquad y + 2 = -\frac{4}{3}$$

just the inside guts!

Subtracting 2 from both sides, we'll write 2 as $\frac{6}{3}$ on the right side of each equation for easier subtraction:

$$y + 2 = \frac{4}{3} \qquad\qquad\qquad y + 2 = -\frac{4}{3}$$

$$y + 2 - 2 = \frac{4}{3} - \frac{6}{3} \qquad\qquad y + 2 - 2 = -\frac{4}{3} - \frac{6}{3}$$

$$y = -\frac{2}{3} \qquad\qquad\qquad y = -\frac{10}{3}$$

Our boundary points are $y = -\frac{10}{3}, -\frac{2}{3}$.

Step 3. Let's graph these. Hmm, where's $-\frac{10}{3}$ on the number line? Let's write it as a mixed number:* $-3\frac{1}{3}$. Much better. At $-3\frac{1}{3}$ and $-\frac{2}{3}$, we'll make the dots *open*, because the inequality symbol was <.

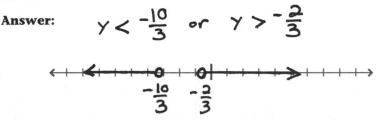

Step 4. Now, will the solution set be the line segment *in between* these values, or will it be the two rays starting at these values and extending *away*? Let's pick an easy point to test between $-\frac{10}{3}$ and $-\frac{2}{3}$; how about $y = -2$? We'll plug it into our inequality from Step 1 and see if it satisfies it: $|y + 2| > \frac{4}{3} \;\rightarrow\; |(-2) + 2| \overset{?}{>} \frac{4}{3} \;\rightarrow\; |0| \overset{?}{>} \frac{4}{3} \;\rightarrow\; 0 \overset{?}{>} \frac{4}{3}$ Nope! So that means our solution set is *not* the segment in between the two boundary points, but rather, it's the two rays extending on either side of the boundary points. Just to double check, let's make sure that a point outside the boundary really does satisfy the inequality. How about $y = 0$? Plugging it in, we get $|y + 2| \overset{?}{>} \frac{4}{3} \;\rightarrow\; |0 + 2| \overset{?}{>} \frac{4}{3} \;\rightarrow\; 2 \overset{?}{>} \frac{4}{3}$ Yep!

Step 5. Let's write our solution set as the disjunction $y < -\frac{10}{3}$ or $y > -\frac{2}{3}$. Done!

Answer:
$$y < \frac{-10}{3} \;\; \text{or} \;\; y > \frac{-2}{3}$$

. . . .

* To review converting improper fractions to mixed numbers and vice versa (remember the MAD face method?), check out Chapter 4 in *Math Doesn't Suck*.

Doing the Math

Solve these inequalities and graph their solution sets. I'll do the first one for you.

1. $-|g + 6| + 7 < 10$

Working out the solution: Remember, we cannot distribute a negative sign to the inside of absolute value bars. To isolate the bars, first we subtract 7 from both sides and get $-|g + 6| < 3$. And then we multiply both sides by -1, so we must flip the direction of the symbol: $|g + 6| > -3$. Hey, this is true for all values of g! (See Mutant Type 3 on p. 112.)

Answer: True for all values of g

2. $|y - 1| \geq 2$

3. $|a - 3| < 2$

4. $-|x| \geq -1$

5. $2|2n + 1| - 1 > 5$ (Hint: Take it one step at a time. First just isolate the bars.)

6. $-4|h - 3| + 1 > 13$

(Answers on p. 404)

What's the Deal: The Mutant Cases

What happens if the "number" after step 1 on p. 107 isn't positive? Oh, so many things can go awry. I like to call these . . . the mutants. In all of these cases, don't be tempted to find "boundary points." They have no meaning and will just confuse you. These mutants can be slippery little suckers.

Mutant Type 1

|something with a variable| < negative number (or zero)

|something with a variable| ≤ negative number

These will have <u>no solution</u>. For example, $|y - 1| < 0$ and $|x + 1| \leq -2$ have no solution. I mean, how can the absolute value of anything (which is a *distance*) be less than zero?

Mutant Type 2

|something with a variable| ≤ zero

The graph of this solution set will be <u>a single closed dot</u>.* Very strange for an inequality, but it *is* a mutant, after all. Consider $|x - 5| \leq 0$. The left side can never be *less* than zero, but it *can* <u>equal</u> zero, which happens at $x = 5$. That's where the dot goes on the number line. Notice that if the symbol were < instead of ≤, there would be no solution. (See above.)

Mutant Type 3

|something with a variable| > negative number

|something with a variable| ≥ negative number (or zero)

These will be true for <u>every real number</u>. For example, $|x + 1| > -2$ is true for all values of x, because no matter what x is, $|x + 1|$ can never be negative, so it will *always* be *greater* than any negative number, including −2. The solution set is <u>the entire number line</u>.

Mutant Type 4

|something with a variable| > zero

This is the strangest of them all. This type of inequality will be true for all real numbers except for one!* For example, $|x + 1| > 0$ is true for all values of x <u>except</u> when $x = -1$. Try plugging stuff in and you'll see what I mean. (Remember, $0 > 0$ is a false statement, but $0 \geq 0$ is a true statement.)

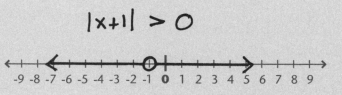

$$|x+1| > 0$$

-9 -8 -7 -6 -5 -4 -3 -2 -1 **0** 1 2 3 4 5 6 7 8 9

* * * * * * * * * * *

* Exceptions would arise if we had, for example, x^2 inside the absolute value bars, which you won't see in Algebra I.

Pretty wacky, huh? Just remember this: After Step 1 from the Step By Step on p. 107, if your "number" is positive, you don't have to even think about this stuff. But if it isn't positive, then my dear, you have a mutant on your hands.

Takeaway Tips

The Mirror rule states that whenever we multiply both sides of an inequality by a negative number, we must reverse the direction of the inequality symbol. This works for synchronized swimmers, too!

Conjunctions are two statements connected with *and*; both must be satisfied. They're very <u>controlling</u> like that. Disjunctions are two statements connected with *or*; either one can be satisfied. They're much more easygoing, even a little <u>distr</u>acted.

Absolute value inequalities will almost always be either a line segment connecting two boundary points (a conjunction) or two rays extending from the boundary points forever in both directions (a disjunction). The exceptions are the mutants!

The symbols < and > mean we'll use empty dots for the boundary points, whereas ≤ and ≥ mean we'll use closed dots for the boundary points.

MOOD ZAPPER:
Take Control of Your Feelings!

We all get bummed out sometimes; life isn't always fair, right? You just failed a test, your best friend isn't acting like a true best friend, and the crush who was so perfect in your imagination turned out to be a jerk in real life. So what can you do? Well, perhaps more than you realize, *you* have the power to feel better.

MOOD ZAPPER 1:
Feeling Better with Music
Put together a playlist of happy songs that always make you feel good. And then you can play them anytime. Music has a powerful effect on our emotions!

MOOD ZAPPER 2:
Feeling Better by Giving Back
The next time you feel sad, angry, or frustrated and can't figure out how to feel better, find a way to HELP someone else, whether by tutoring a fellow student or even helping your parents around the house. Imagine their surprise! Witnessing the power you have to make others feel good puts you back in the driver's seat of your emotions.

MOOD ZAPPER 3:
Feeling Better by Getting Off the Judge's Bench
Most of us witness "injustice" all around us, and it can eat us up inside. But think about that horrible bully or the popular girl who seems to enjoy making other girls' lives miserable. You might feel tempted to get back at him/her somehow, and bring justice to the world. But that's not your job! For all you know, she'll end up having a heart attack from guilt and stress later in life, or she'll always feel a dull aching insecurity that she'd never, ever admit. It's actually sad. The point is, it's not up to *you* to make sure justice is served. The universe has a way of doing that for us, so you truly can let it go. The sooner you realize that, the happier you'll be.

MOOD ZAPPER 4:
Feeling Better Through Exercise
Seriously, for a quick fix, get up and do 20 jumping jacks, go for a run, or close your door, put on your favorite playlist, and dance. With energy rushing through your body, you can get a new perspective that might help you focus on studying for that upcoming test or have a little more patience when dealing with your family. Our emotions are chemical, and you can literally change the chemicals running through your body by just jumping up and down. Try it!

And here's what YOU had to say about it:

If I'm in a really bad mood, I think, "Will I even remember [insert what I'm upset about here] in a month's time?" Looking at the big picture can usually cheer me up. **Victoria, 14**

I can brighten my mood just by thinking about silly things like a clown hitting himself. **David, 16**

When I am upset about something, I just take a deep breath and think about doing something that I like to do and pretend I'm there doing it. **Dominique, 16**

Sometimes I can lift my mood by drawing in my sketchbook. It helps me get away from everything and calm down. **Christopher, 16**

I used to blame my mother for putting me in a bad mood, because of an argument or something, but the truth is, she didn't put me in that mood; I did! I realized I had the control to decide what was going to make me angry. It's all a matter of perspective. **Jade, 15**

It makes me feel happy when I make other people happy. I also like to get hugs. **Emily Mae, 16**

If I wake up on the wrong side of the bed one morning, I just look forward to volleyball. If I'm still upset, I let myself be mad for one day. The next day, I have to shape up and realize it's not the end of the world. And God is always with me, guiding me through. **Chloe, 13**

Look, being in a bad mood isn't the end of the world, but it's important not to stay in it too *long*. Isn't it nice to know we have that power? For more ideas, check out pp. 366–67.

Chapter 9

Can You Keep a Secret?

Functions and Relations

*I*f you like secrets, you're in for a treat: This chapter is full of whispers, secret recipes, forbidden ingredients, and trying to figure out who you can trust. And where does all this intrigue take place? Why, a sausage factory, of course.

Functions and Sausage Factories

Here are some equations for **functions**:

$$f(x) = 2x + 1 \qquad f(x) = \frac{x}{2} - 3 \qquad f(x) = 3x^2 - 1$$

Functions are often in the form "$f(x) = $ something involving x," and they're a lot like factories: They take IN a value of x and put OUT some other value, which we call $f(x)$. More specifically, if we think of functions as little sausage factories, then we can think of the x values as the ingredients and the $f(x)$ values as the sausages. An ingredient could be chicken, beef, tofu, and so on, but no matter what ingredients the factory gets, it will do the same things to them every time, just like functions do the *same things* to the input numbers they get:

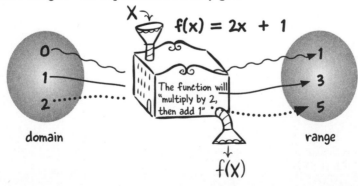

For the function $f(x) = 2x + 1$, no matter what ingredient (value of x) we put in, the factory will always double the ingredient and then add 1 to it. For $f(x) = \frac{x}{2} - 3$, the factory will take the ingredient, cut it in half, and then subtract 3 from it. That's not exactly how I like my chicken prepared, but whatever.

Notice that we can figure out as many ingredient/sausage combos as we want by simply plugging input values into the function's equation and getting the output values.* For the rest of this chapter, however, we're going to focus on functions defined in a more abstract way; they don't even *have* equations. Yep, some functions are defined strictly by their ingredient/sausage pairings, often with "mapping diagrams" like the one below. *How* we get from input to output might remain a mystery, and that's fine! Hey, some factories are more private than others, what do you want?

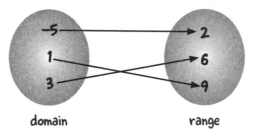

domain   range

In fact, you could make your own drawing like this, put some numbers in the bubbles, draw arrows between all of them, and say, "Look, I made a function."[†] Just make sure you obey this function rule: An input value (ingredient) can't be paired with two different output values (sausages). On the other hand, more than one ingredient *is* allowed to result in the same sausage. Think of it this way: In the mapping diagram of a function, it needs to be clear where everyone is *going*.

· · · · · · · · · · ·

* For tons of practice doing this, see Chapter 17 in *Kiss My Math*.
† A proud moment.

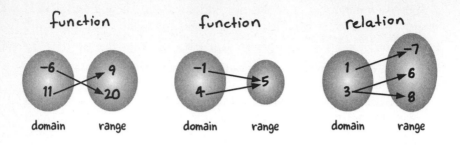

For the two functions above, even though the arrows cross and overlap, each ingredient knows where it's *going*. Hmm, 5 seems to be a popular destination, doesn't it? However, the mapping diagram on the right is *not* a function because I mean, where is that 3 going? It looks like it can't make up its mind. So we have a **relation**, not a function. You see, like factories, functions should be trustworthy and dependable. If we put in 3, we shouldn't have to wonder if we'll end up with 6 or 8. When we put a particular input in, we should know with *confidence* what we'll get.

Notice that the set of input values is called the **domain** and that the set of output values is called the **range**. Let's straighten out all this function vocabulary!

What Are They Called?

Function, Relation, Domain, and Range

Function

A **function** pairs input values with output values, often defined by an equation like $f(x) = x + 2$ or $f(x) = 3x^2 - 1$. But sometimes they are defined *without* an equation, as simply a collection of pairs of numbers (with bubbles, or as ordered pairs: see above and p. 120). Functions are a special type of **relation**—the trustworthy type: Each input value gives only *one* output value.

Relation

A **relation**, however, can match more than one output for each input. So we might not always know what an input's result will be. For example, as on p. 118, when 3 is the input, we don't know if we'll get 6 or 8 as the output. Yikes! As with functions, relations can be defined by an equation,* with a mapping diagram, or as a set of ordered pairs.

Domain

The **domain** is the set of input values for a function (or relation). In other words, the domain is the full list of *all possible ingredients*. For functions (or relations) like the ones on p. 118, the domain is the set of values in its first bubble. After all, in that particular factory, those are the only possible ingredients we know how to get sausages from.

Range

The **range** of a function (or relation) is the set of possible output values—all possible sausages. For functions (or relations) like the ones on p. 118, the *range* is just the set of values in its second bubble. After all, those are the only sausages we can get.

We'll see other examples of domains and ranges on pp. 125–26.

QUICK NOTE   Keeping it straight: The *domain* is the list of ingredients (inputs), and the *range* is the list of resulting sausages (outputs). Think about the phrase "the main ingredients," like how the main ingredients in many cake frostings are sugar and butter.[†] Now pretend you're a baby who can't pronounce "the" and it becomes "da main ingredients," right? And voilà! Domain = ingredients.

• • • • • • • • • •

* In case you're curious, here's an example of an equation for a **relation** that is not a function: $x^2 + y^2 = 9$. It's a circle with a radius of 3. More on that in Algebra II!
[†] There can be others, like vanilla extract, but sugar and butter are the *main* ingredients.

Ordered Pairs and Whispering Parentheses

Whenever I see something written in parentheses, I think of someone whispering. (See what I mean? You can make your "inside reading voice" sound like a whisper!)* Whispering is good for telling secrets, don't you agree?

We've already seen how a function can be defined just by listing the inputs and outputs in bubbles with arrows in between. But we can also do this in a much more efficient and whispery way . . . for those top-secret sausage recipes.

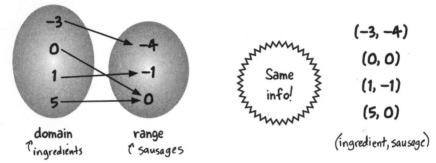

domain
↑ingredients

range
↑sausages

Same info!

(−3, −4)
(0, 0)
(1, −1)
(5, 0)

(ingredient, sausage)

Ordered pairs give *all the same information* as mapping diagrams. So notice that just by looking at any set of ordered pairs—for example, (2, 3), (1, 3), and (0, −1)—we know it could belong to a function because if I tell you an ingredient, you could say with certainty which sausage will result. However, (2, 3), (2, 1), and (0, −1) could *not* belong to a function, because if I told you the ingredient 2, you wouldn't be able tell me if the sausage would be a 3 or a 1. And that's just not very factory-like.

Shortcut Alert

When looking at the ordered pairs of a relation, we can tell if it's a function or not with this simple trick: If the same *ingredient* happens more than once, it can't be a function! The only exception is if someone tries to play a trick on you

* * * * * * * * * *

* You might remember the secret sausages and whispering parentheses from Chapter 17 of *Kiss My Math!*

and gives you a list like this: (2, 3), (5, 7), and (2, 3). See, this *could* represent a function, because even though the 2 shows up as an ingredient twice, it's paired with the same sausage, 3, both times! So it's still a reliable, trustworthy function: When you stick in 2, you'll always get a 3.

QUICK NOTE   SET NOTATION: When dealing with a "set of values" or a "set of ordered pairs," we can use those squiggly parentheses as set notation, which is just a way to indicate that we've collected them all together. So instead of saying, "The domain is 2, 3, and 4," we could say, "The domain is {2, 3, 4}." And instead of saying "The set of ordered pairs for this function is (0, 1), (2, 8), and (4, 5)," we could say, "The set of ordered pairs is {(0, 1), (2, 8), (4, 5)}." No big deal, right? We're just collecting them in squiggles.

Watch Out!

When looking for repeating *inputs*, don't get confused by the outputs! For example, the relation {(4, 4), (5, 5), (6, 5)} is totally a function: Each ingredient **has exactly** one output assigned to it. However, {(4, 4), (5, 5), (5, 6)} is NOT a function because the input value 5 has two possible outputs: 5 and 6. I guess we have to pay attention . . .

QUICK NOTE   In bubble "mapping diagrams," numbers from top to bottom always go in order from least to greatest; that's why the arrows sometimes have to cross each other.

Doing the Math

Determine if each set of ordered pairs defines a function or just a relation. Name its domain and range, and draw its bubble diagram with arrows. I'll do the first one for you.

1. {(0, 2), (–9, 2), (2, 2)}

<u>Working out the solution</u>: Could these ordered pairs define a function? Yep! None of the input values repeat. (It's okay that the output values repeat, though.) To draw the bubbles, we'll put all the "ingredients," the domain values, on the left and draw arrows to their "sausages," the range values. Let's put them in order and use set notation.

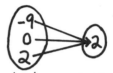

Answer: It's a function; the domain is {–9, 0, 2}; the range is {2}.

2. {(0, 0), (2, 2), (3, –3)}

3. {(–1, 9), (–1, 3), (–1, 10), (–1, 1)}

4. {(0, 8), (3, –4), (–7, –4), (5, 8)}

5. {(3, 6), (2, 6), (3, 6)} *(Be careful!)*

(Answers on p. 404)

Danica's Diary
TOP SECRET JOURNAL WRITING

Look, we all have thoughts that are so secret, the only place we can express them is in a journal. But what if the mere idea of someone finding your diary is enough to keep those thoughts totally bottled up inside? It could be that you have a huge crush on someone who must never find out, or maybe you're so angry at someone you actually feel like hurting the person. No matter how shameful or evil or "unlike you" your thoughts might be, it can really help to get them off your chest. I know this firsthand, so I'm going to share a trick I came up with when I was a teenager. Try this: Take a piece of paper, and start writing on the top line. Then, instead of moving to the next line, *keep writing* on top of what you just wrote. Soon, no one will be able to read your thoughts, not even you. Then, if the feelings you're venting are really angry or evil, you'll find that you start to calm down, because you got them out of your system, but you don't have to look at them! And this part is important: As you begin to feel calmer, *keep writing* on that same line, saying things like "I have these feelings, but I won't do anything bad with them because . . ." and fill in the blanks. Keep writing on that line until you find the "you" that you are proud of, even if it takes awhile. You'll feel so much better, and you'll have learned a lot about the real you. I'm telling ya, it really works. To hear how other young women use writing in their lives, check out p. 21 and pp. 255–56!

Sausage Factories on a Plane

I bet it would be really expensive to put a whole sausage factory on an airplane and move it to a different city. So I'm really glad we don't have to do that.

Instead, we're going to put these sausage factories on a different kind of plane—the coordinate plane!* At this point in your life, you've plotted many points. Well, "whispering" parentheses of ordered pairs, like (–2, 2), (3, 2), (4, –3), can indeed be plotted as points on the coordinate plane. Check out these three ways to express the *same* function:

. . . .

* Some textbooks call this the Cartesian plane.

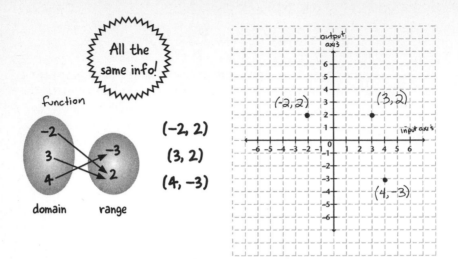

Now check out this relation that is *not* a function:

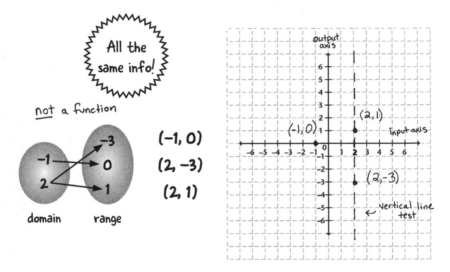

Notice on the above graph how we get points stacked up on top of each other. That's how we can tell it's not a function, because an input value (ingredient) has more than one output. See the vertical line I drew? This is often called the vertical* line test.

• • • • • • • • • • •

* Vertical means "up and down," which is different from horizontal, which means "side to side." Horizontal is easy to remember because of sunsets on the horizon!

The Vertical Line Test

When looking at the graph of a relation, if an imaginary vertical line could ever pass through more than one point at a time, then oops! It's not a function.

Notice on the first graph on p. 124 that no vertical line could ever pass through more than one point. That's because it *is* a function.

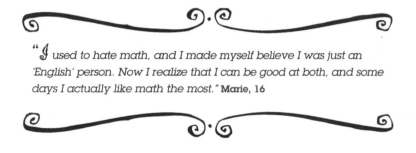

"*I* used to hate math, and I made myself believe I was just an 'English' person. Now I realize that I can be good at both, and some days I actually like math the most." **Marie, 16**

Domains and Ranges . . . Again

A picture really is worth a thousand words. Just by looking at a graph, we can figure out if we're looking at a function or not. We can also determine its *domain* and *range*. (Reread their definitions on p. 119!) To find the domain and range, we ask: What values of *x* are used as ingredients? What values of *y* are used as resulting sausages?

In order to figure out the domain from a graph, I like to run my finger back and forth along the *x*-axis to check <u>which *x*-values were used as ingredients</u>. Remember that all those little tics on the flat, horizontal axis represent *x*-values* (the ingredients). So, for each *x*-value you put your finger on, stop and then trace your finger directly up (or down). If your finger hits a graphed point, that means the *x*-value is indeed an ingredient; in other words, it's in the domain.

. . . .

* And, of course, there are infinitely many *x*-values between each tic mark, too!

The same goes for ranges. Run your finger up and down on the tics of the vertical axis. For each *y*-value, trace your finger side to side, and if it hits a graphed point, then that sausage was indeed made at this factory. In other words, that *y*-value is in the range.

For example, in Graph A, we can see that <u>all</u> *x*-axis values are used as ingredients in the function. (Remember that the arrows on the function indicate that the lines keep going forever.) No matter where our finger is along the *x*-axis, we can always trace our finger directly up and hit the line. However, it appears there are no negative *y*-values graphed, so the range is just **y ≥ 0**. Make sense? As you know, when we say $y \geq 0$, we're including all the values of *y* that are positive and also zero.

In Graph B, all *x*-axis values are used and all *y*-axis values are attained, so the domain and range are both <u>the set of all real numbers</u>.

Check it out: Graph C doesn't pass the vertical line test, so it's not a function. It seems that no negative *x*-values are ever used, just zero and all positive values of *x*, but all *y*-axis values *are* achieved. So the domain is **x ≥ 0**, and the range is the set of all real numbers.*

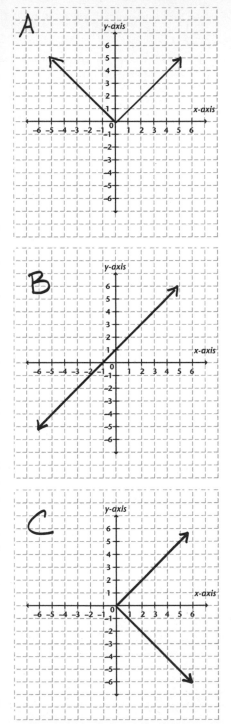

• • • • • • • • • •

* If you're curious, Graph A is $y = |x|$, Graph B is $y = x + 1$, and Graph C is $|y| = x$.

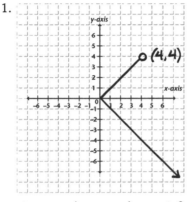

Determine if each graph represents a function or just a relation, and state its domain and range. I'll do the first one for you.

1.

<u>Working out the solution</u>:
This graph does not pass the vertical line test, so it's not a function; it's just a relation. What's the domain? Running a finger along the x-axis shows there are no negative x-values with corresponding y-values. The smallest such x-value input is zero. All the positive x-values are also used for this relation, because for any of them, we can trace our finger down and hit graphed points! So the domain is **x ≥ 0**. Similarly, the range includes all negative values, but at y = 4, there is no graph—only an empty dot. And no y-values bigger than 4 are graphed, either. So the range is **y < 4.**

Answer: A relation; the domain is x ≥ 0, and the range is y < 4.

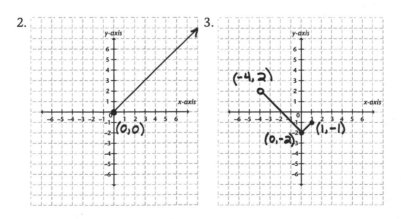

2. (0,0)

3. (-4, 2) (0,-2) (1,-1)

4.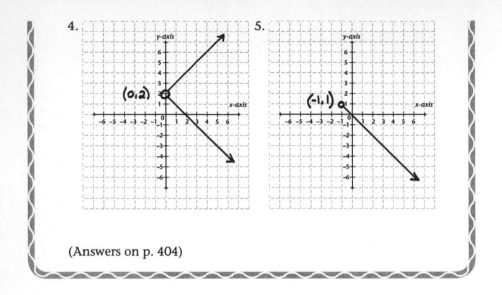

5.

(Answers on p. 404)

This stuff can take some getting used to, but believe me, it's time well spent. Teachers and standardized test writers just love these kinds of problems.

QUICK NOTE   In the next chapter, we'll graph functions defined by equations. But instead of writing $f(x)$, we'll just write y. For example, these are the same function, written differently:

$$f(x) = x + 1 \qquad y = x + 1$$

It's also more familiar to label the up-and-down axis as the y-axis. I've always thought that y looks a bit more like a sausage than $f(x)$ does anyway.

Takeaway Tips

 A function is a relation with one extra rule: Input values must have only *one* output each. In other words, when we put an ingredient into a function, we should be able to say with certainty which sausage we will get.

 The domain is the set of all allowed input values, and the range is the set of all resulting output values. Remember: da main ingredients → domain = ingredients!

 The vertical line test can determine if a graph is a function or just a relation.

 To find the domain and range of a relation or function from a graph, use your finger to run across *x*-values or *y*-values to see if there is a corresponding point on the graph for it.

Chapter 10

They're Going Steady

Equations of Lines, and Graphing Lines and Linear Inequalities

$\mathcal{H}$ave you started dating yet? I hate to say it, but it's not all it's cracked up to be. I mean, if you stay true to yourself, you'll eventually find a guy who totally respects and supports your dreams and goals and also gives you butterflies. Until then? Let's just say you'll feel plenty of butterflies, but they might come from frogs. For instance, if the guy ever tries to make you feel bad about yourself, then no matter how cute he is, you've got better things to do with your time, sister!

It might be tempting to see other people in steady relationships and wish you had what they do. But remember, you can't ever really know what their relationship is like the rest of the time. On the other hand, we *do* know what the relationship is between *x* and *y* for every point on a line. (You knew I'd get around to the math, didn't you?)

Linear Functions and Steady Relationships

In the previous chapter, we saw a few functions defined by equations, and here are some more:

$$y = 3x \qquad y = \frac{2}{3}x - 4 \qquad y = -6x + 2$$

These particular functions all have something in common: They are *linear* functions, so if we graphed them, they'd look like straight lines. Another way to think about this is that they each describe a steady, *linear relationship* between *x* and *y*.

For example, on the graph of *y* = 3*x*, every single point (*x*, *y*) shares the exact same relationship between its *x*- and *y*-value: The *y*-value will be *three times* as big as its corresponding *x*-value. In fact, here are some points on this line: (1, 3), (3, 9), (–2, –6). See? The coordinates are all tied together by the same, steady, linear *relationship*.

In fact, any (x, y) combo that you can think of, where the y-value is three times the x-value, *must* be a point that falls on the line $y = 3x$, even something crazy like (300, 900). If our graph paper were big enough, we'd see that yep, it's on the line.

Let's take a moment to briefly review the basics of lines with a lightning-fast refresher from pre-algebra.*

A Quick Review of the Slope-Intercept Form of Lines

Equations for lines are often written in something called **slope-intercept form**:

$$y = mx + b$$

. . . where m and b are real numbers, and x and y are the two variables that share a linear relationship described by the equation. The m is called the **slope**, and it describes the steepness of the line. The m makes me think of the side of a <u>m</u>ountain where you might find a ski slope. The slope of a line is often written as a fraction, $\frac{rise}{run}$, where the *rise* is the change in y-value and the *run* is the change in x-value. Sound familiar?

The b is called the **y-intercept**—the spot on the y-axis where the line crosses it, so the point (0, b) will always be on the line. Think of b as the <u>b</u>lade of a sword, striking the tall y-axis. That's pretty much how the line looks when it intercepts the y-axis. The only lines that don't have a y-intercept are vertical lines (like the line $x = 5$).

* * * * * * * * * *

* For a full review of plotting points and graphing lines, including the bubblegum trick, check out Chapter 18 of *Kiss My Math*. I highly recommend reviewing that chapter *any* time you are about to study lines in algebra and beyond. It's a good, solid base, with tons of easy tricks for remembering stuff!

Consider the line $y = \frac{1}{2}x - 1$.

The slope is $\frac{1}{2}$, and the y-intercept is **–1**. To graph the line, we can just pick some easy x-values, like $x = 0, 2, 4$, and plug them in to find their corresponding y-values. (Plugging in $x = 0$ will always give us the y-intercept.) We get:

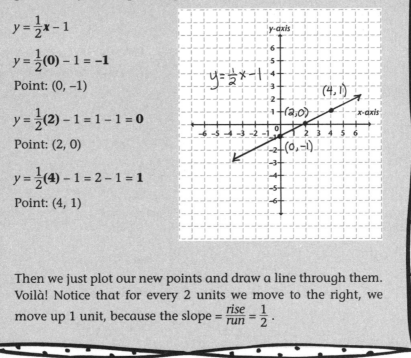

$y = \frac{1}{2}x - 1$

$y = \frac{1}{2}(0) - 1 = \mathbf{-1}$

Point: (0, –1)

$y = \frac{1}{2}(2) - 1 = 1 - 1 = \mathbf{0}$

Point: (2, 0)

$y = \frac{1}{2}(4) - 1 = 2 - 1 = \mathbf{1}$

Point: (4, 1)

Then we just plot our new points and draw a line through them. Voilà! Notice that for every 2 units we move to the right, we move up 1 unit, because the slope $= \frac{rise}{run} = \frac{1}{2}$.

Slope-Intercept Form vs. Standard Form: Drawing the Battle Lines

So, we've all graphed lines in slope-intercept form, $y = mx + b$, but there's another common way to write linear equations. It's called **standard form**. Which form is more useful? Which will reign champion? The battle "lines" have been drawn. . . .

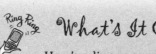

What's It Called?

Here's a linear equation in **standard form**:

$$Ax + By = C$$

...where *A*, *B*, and *C* are integers* with no factors in common, *A* > 0, and *B* ≠ 0 (in other words, *A* and *B* can't be zero, and *A* can't be negative, either).[†]

Here are some linear equations in *standard form*.

$$x + 7y = 0 \qquad 6x - 5y = -8$$

But the linear equations below are *not* in standard form. (The first one has a non-integer coefficient, and the second one has common factors.)

$$\frac{x}{2} + 7y = 0 \qquad 6x - 4y = -8$$

FYI, the following are *not linear* equations. Their graphs wouldn't even be straight lines.

$$xy = 9 \qquad \frac{1}{x} + 2y = 5 \qquad 6x^2 - 5y = -8$$

In order to write a line in standard form, $Ax + By = C$, we just subtract or add terms to both sides so *x* and *y* are on the *same side* of the equation, and multiply or divide both sides by whatever we need to make sure there are no fractions or common factors.

For example, to write $y = \frac{3}{4}x + 5$ in standard form, we'd subtract $\frac{3}{4}x$ from both sides: $-\frac{3}{4}x + y = 5$, and then we'd multiply both sides by 4 and get $\mathbf{4}\left(-\frac{3}{4}x + y\right) = \mathbf{4}(5) \rightarrow -3x + 4y = 20$. The first term shouldn't be negative, so we'd also multiply both entire sides by –1 and get the answer: $\mathbf{3x - 4y = -20}$.

. . . .

* No fractions or decimals allowed. For a formal definition of *integers*, see p. 398.

[†] Advanced footnote: The restriction that A can't be zero is part of the formal definition of *standard form*, but notice that, for example, $y = 3$ is indeed a line (with $A = 0$, $B = 1$, $C = 3$); it's horizontal. So this definition just means that horizontal lines can't be written in standard form. No big deal. Vertical lines, like $x = 5$, can't be written in standard form *or* slope-intercept form.

Have you decided which form you like more, standard or slope-intercept? If you *really* hate fractions, you might prefer standard form. But you'll see that slope-intercept form is so much easier to work with because, just by looking at it, we can *see* the slope and the y-intercept, which makes it easier to graph and do all sorts of other things. I'll be honest. I don't like standard form very much. Okay? There. I said it.

QUICK NOTE   Whenever we're asked to do something with a linear equation in standard form, I recommend rewriting it in slope-intercept form, $y = mx + b$. It makes life so much easier.

Say we're asked to find the slope of this line: $2x + 3y = 21$. Seriously, the fastest way to find its slope is to write it in slope-intercept form. And how do we do that? We just solve for y!* Isolating y, first we'll subtract $2x$ from both sides, and then we'll divide both entire sides by 3:

$$2x + 3y = 21$$

$$\rightarrow \quad 3y = -2x + 21$$

$$\rightarrow \quad \frac{3y}{3} = \frac{-2x + 21}{3}$$

$$\rightarrow \quad y = \frac{-2x}{3} + \frac{21}{3} \,^\dagger$$

$$\rightarrow \quad y = -\frac{2}{3}x + 7$$

And voilà! Now we can easily see that the slope is $-\frac{2}{3}$. As a bonus, we've also discovered that the y-intercept is 7.

. . . .

* Remember solving for one variable in terms of another from pp. 78–80?
† Notice we can "split" the fraction, because it's just fraction addition in reverse. Since $\frac{-2x}{3} + \frac{21}{3} = \frac{-2x + 21}{3}$, it's also true that $\frac{-2x + 21}{3} = \frac{-2x}{3} + \frac{21}{3}$. Clever, right?

In order to rewrite an equation into slope-intercept form, $y = mx + b$, just <u>solve for y</u> in terms of x.

Are They in a Relationship?
Verifying that a Point Is on a Line

So, someone gives you an equation for a line and a point and says, "Is this point on this line?" That might sound hard, but it's one of the easiest things to do with a line. This is the same as asking, "Does this point have the *relationship* described by this line?" We just plug the point into the equation for the line and see if we get a true statement!

Step By Step

Verifying that a point is on a line:

Step 1. Plug the *x*-value and *y*-value from the point (x, y) into the equation of the line.

Step 2. Simplify. If you get a true statement, then the point is on the line. If you get a false statement, then the point is not on the line. That's it!

And... Action! Step By Step In Action

Is the point (4, 3) on the line $5x - 4y = 8$?

Steps 1 and 2. Let's plug $x = 4$ and $y = 3$ into the equation and simplify:

$$5(4) - 4(3) \overset{?}{=} 8$$

$$20 - 12 \overset{?}{=} 8$$

$$8 = 8 \quad \text{Yep!}$$

We got a true statement, so the point (4, 3) is indeed on the line. Awesome.

Watch Out!

When plugging a point into the equation for a line, be sure to plug in the x-coordinate for x and the y-coordinate for y, and not the other way around. It can be tempting to reverse them, especially when y comes first in the equation, as it does in slope-intercept form. For example, if you were going to plug the point **(3, 5)** into the equation **$y = 2x - 1$**, it would be easy to do this by mistake: $3 \overset{?}{=} 2(5) - 1 \rightarrow 3 \overset{?}{=} 10 - 1 \rightarrow 3 \neq 9$ . . . and then you'd think the point isn't on the line, when in fact it is. Yikes! Doing it correctly, we get $5 \overset{?}{=} 2(3) - 1 \rightarrow 5 \overset{?}{=} 6 - 1 \rightarrow 5 = 5$. Yep! Just pay close attention: When you're plugging in the values, point to the coordinate numbers in the parentheses with your pencil and think to yourself *this is x and this is y*, and you'll be good to go.

Just like we have a y-intercept, there's also something called the **x-intercept**, and it's just what you'd expect!

Ring Ring What's It Called?

The **x-intercept** is the number on the x-axis where the line crosses it. This happens when $y = 0$. So to find the x-intercept, just plug in 0 for y, and solve for x. When you plot this point, you'll see that it lands on the x-axis. The only lines that don't have an x-intercept are horizontal lines (like the line $y = -3$).

For example, to find the x-intercept of $5x + y = 20$, we plug in 0 for y and get $5x + \mathbf{0} = 20 \rightarrow \mathbf{x = 4}$. And that's where the line crosses the x-axis, at (4, 0).

QUICK (REMINDER) NOTE When plotting points to draw the graph of a line, even though we only *need* two points, it cuts down on mistakes when we use three points.

The exercises below will combine stuff from this entire chapter so far, so flip back if you need to check on how something is done.

Doing the Math

For each linear equation:

a. Determine if the given point is on the line.

b. Write it in standard form if it isn't already.

c. Write it in slope-intercept form, $y = mx + b$, by solving for *y*.

d. Find the slope, *y*-intercept, and *x*-intercept of the line.

e. Graph it by plotting three points of your choice.* I'll do the first one for you.

1. $\frac{y}{2} + \frac{x}{3} = 1$; point (5, 4)

<u>Working out the solution</u>: For part **a**, we'll just plug in the point and see if we get a true statement: $\frac{4}{2} + \frac{5}{3} \overset{?}{=} 1 \rightarrow 2 + \frac{5}{3} \overset{?}{=} 1$ Nope! The point is *not* on the line. For part **b**, we multiply both sides of the original equation by 6 and get $6\left(\frac{y}{2} + \frac{x}{3}\right) = 6(1) \rightarrow$ $3y + 2x = 6 \rightarrow 2x + 3y = 6.$ For part **c**, we now subtract 2x from both sides and divide both sides by 3: $y = -\frac{2}{3}x + 2.$

. . . .

* The graphing part should be review. If not, check out Chapter 18 in *Kiss My Math*.

For part **d**, we can now see that the slope is $-\frac{2}{3}$ and the y-intercept is **2**, which is at the point (0, 2). For the x-intercept, we'll plug in 0 for y into the original equation and solve for x: $\frac{0}{2} + \frac{x}{3} = 1 \rightarrow \frac{x}{3} = 1 \rightarrow x = 3$. The x-intercept is **3**, which is at the point (3, 0). Part **e**: Let's plot the x- and y-intercepts **(0, 2)** and **(3, 0)**, and find one more point. Hmm, how about x = 6? We get $y = -\frac{2}{3}(6) + 2 \rightarrow y = -\frac{12}{3} + 2 \rightarrow y = -4 + 2 \rightarrow y = -2$. So a third point to plot is **(6, −2)**. We draw a straight line through them, and we're done. (Pant, pant!)

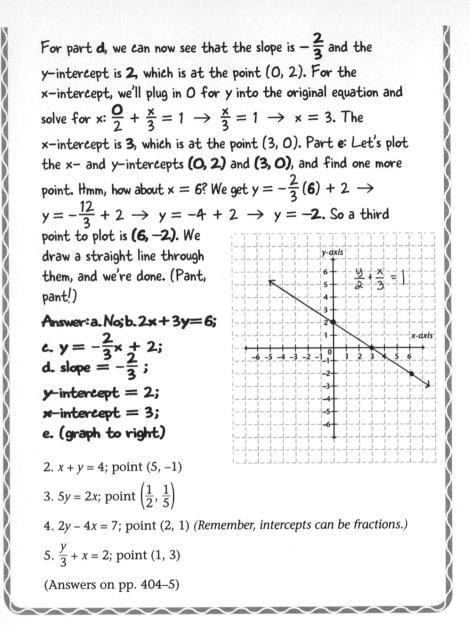

Answer: a. No; b. $2x + 3y = 6$;

c. $y = -\frac{2}{3}x + 2$;

d. slope $= -\frac{2}{3}$;

y-intercept = 2;

x-intercept = 3;

e. (graph to right)

2. $x + y = 4$; point (5, −1)

3. $5y = 2x$; point $\left(\frac{1}{2}, \frac{1}{5}\right)$

4. $2y − 4x = 7$; point (2, 1) *(Remember, intercepts can be fractions.)*

5. $\frac{y}{3} + x = 2$; point (1, 3)

(Answers on pp. 404–5)

One-Sided Relationships: A Note About Horizontal and Vertical Lines

$\mathcal{S}$ome relationships are so one-sided; one person expects all the attention, y'know? Some equations for lines are totally one-sided, too.

For example, when the slope of a line is zero ($m = 0$), it makes the mx term disappear, and we get an equation like **$y = 3$**. Yep, that's an equation for a line; this is a horizontal (flat) line of constant height "3." Some points on it are (0, **3**), (–1, **3**), and (20, **3**). Equations for lines describe a *relationship* between the *x*- and *y*-values. But I mean, this one doesn't even care what the *x*-values are; *y* has decided to always be 3, no matter what. Talk about self-centered.

Similarly, **$x = 4$** is the equation for the vertical line (goes straight up and down), crossing the *x*-axis at 4, with points like (**4**, 0), (**4**, –1), and (**4**, 79). It doesn't care at all about *y*. The slope of all vertical lines is bigger than any number (this is infinity), so it's called "undefined." *Zero* and *undefined* are fitting slopes for these "uncaring" lines, aren't they?

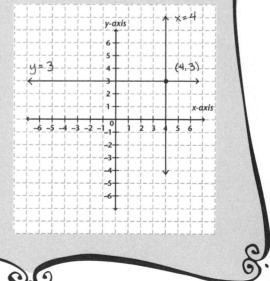

By the way, guess what point they cross at? The only one that satisfies both of their requirements: (4, 3).

Danica's Diary
BREAKING UP IS HARD TO DO

When I was 15, I experienced my first breakup. I was dating a guy named Jeremy, my first "puppy love" boyfriend. I had my first off-screen kiss with him,* and things stayed really sweet (only closed-mouth kissing!). But he was younger by a year, and after eight or nine months, I started seeing him as a friend and nothing more. I still cared about him though and dreaded breaking up. But I made a decision to do something that took a lot of courage: to break up with him in person. I mean, I could have taken the easy way out and done it over the phone, or worse, by writing him a note, but I knew this was the right thing to do. I mean, who wants to be broken up with in a text or email?

I remember being driven to his house; my heart was pounding. When he answered the door, he gave me flowers! I started crying right then. I told him how my feelings had changed. He was so understanding and cool about it. To this day, he has told me that he always respected how I handled things.

Every time we do the right thing, especially when it's hard, we earn respect for ourselves, so that we can look ourselves in the mirror and say, "I'm proud to be me." It's so important, and it applies not only to relationships but also to schoolwork and just about anything else you can think of. In fact, struggling through math problems is an excellent way to practice doing "hard things." The confidence and stamina you get from conquering difficult concepts in algebra can teach you how to conquer *anything* in life, even breakups, and build up your self-esteem, too!

. . . .

* At 12, I had my first kiss "on-screen" in *The Wonder Years,* but that doesn't count!

Graphing Linear Inequalities

Remember how we graphed inequalities in Chapter 8, like $x > 3$ and $w \leq -2$? Well, when inequalities have *two* variables in them—just like lines—then they are called *linear inequalities*. And in fact, graphing linear inequalities is just like graphing lines, really! The only difference is that sometimes we'll use a dotted line instead of a solid line, and get your crayons out, because we've got some shading to do.

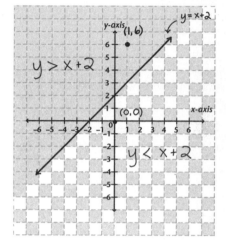

First of all, notice that the line $y = x + 2$ divides the coordinate plane into *three regions*: the top shaded area where all the points have the relationship $y > x + 2$, the skinny line where all the points have the relationship $y = x + 2$, and the bottom checkered area where all the points have the relationship $y < x + 2$.

Look at the top region: $y > x + 2$. Every single point in the shaded region *above* that line will satisfy this inequality, like for example (1, 6). Plugging that point in, we get:

$$y > x + 2$$
$$6 \overset{?}{>} 1 + 2$$
$$\rightarrow 6 \overset{?}{>} 3 \quad \text{Yep!}$$

On the other hand, any point in the *bottom* checkered region will satisfy the other inequality, $y < x + 2$. For example, plugging the point (0, 0) into this bottom inequality, we get:

$$y < x + 2$$
$$0 \overset{?}{<} 0 + 2$$
$$\rightarrow 0 \overset{?}{<} 2 \quad \text{Yep!}$$

As it turns out, checking points like this is an excellent way to figure out which side of the line to shade when we're graphing inequalities. By the way, I used a checkered region here only because we graphed *two* inequalities (and a line) all on the same graph. In typical problems, we'll just shade the side we're graphing, and the other side will stay white.

Graphing linear inequalities:

Step 1. If the inequality symbol is ≤ or ≥, you'll be using a solid line, because the graphed solution (your shaded region) will include all the points that are actually on the line. If it's < or >, you'll be using a dotted line: Your shaded region will *not* include any points on the line. Make a mental note!

Step 2. Temporarily replace the inequality symbol with an equals sign: Now you have an equation for a line. Graph this line as you normally would.* Remember to use a dotted line if the inequality was < or >.

Step 3. Time to shade! To figure out which side to shade, pick a point that is clearly on one side of the line you've graphed. (Important: This cannot be a point that is actually on the line.) Plug that point into the ORIGINAL inequality you started out with. If you get a true statement, then shade the entire region that holds the point you tested. If you get a false statement, shade the other side. Label your graph with the original inequality. Done!

And... Step By Step In Action
Action!

Graph the inequality $\frac{y}{2} + \frac{x}{3} \leq 1$**.**

This looks wacky. Um, what the heck does it mean again? Ah yes, we want to graph all the points in the plane that *satisfy* this inequality—all the (x, y) pairs that make it a true statement. Let's use the step-by-step method:

Step 1. The inequality symbol is ≤, so we'll use a solid line.

Step 2. Replacing the ≤ symbol with =, we get the equation for the line: $\frac{y}{2} + \frac{x}{3} = 1$. On pp. 137–38, we rewrote this as $y = -\frac{2}{3}x + 2$ and graphed it by finding the points (0, 2), (3, 0), and (6, –2).

. . . .

* We've done some graphing in this chapter, but for a full review of graphing lines, see Chapter 18 in *Kiss My Math* if you need it.

Step 3. Notice that the point (0, 0) is clearly not on the line, so it's a safe point to use for plugging. Let's see if that side of the plane is the one we shade or not. Plugging that point into our original inequality, we get:

$$\frac{y}{2} + \frac{x}{3} \le 1$$

$$\rightarrow \frac{0}{2} + \frac{0}{3} \overset{?}{\le} 1$$

$$\rightarrow 0 + 0 \overset{?}{\le} 1 \text{ Yep!}$$

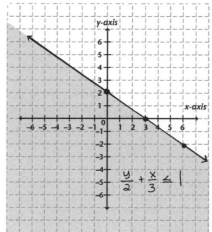

So we shade the side of the line that includes the point (0, 0), which in this case is the entire region *below* the line. And we're done!

Watch Out!

When picking points to check for shading, don't pick points that are actually *on* the line. They won't help us figure out how to shade, because points on the line will *always* satisfy the inequality if the symbol is ≤ or ≥, and points on the line will *never* satisfy the inequality if the symbol is < or >. And that's not very helpful, now, is it?

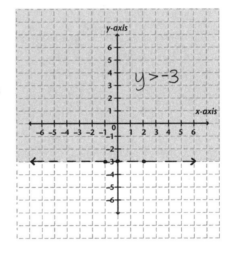

Doing the Math

Graph these inequalities on the coordinate plane. I'll do the first one for you.

1. $y > -3$

<u>Working out the solution</u>: We see >, so we know we'll use a dotted line. We start by graphing the line $y = -3$. Wait, where's the x? Ah, this is one of those one-sided relationships we talked about on p. 139. No matter what x is, y will always be -3, so let's plot points like $(0, -3)$, $(-1, -3)$, and $(2, -3)$ and draw our dotted line through them. Let's pick the point $(0, 0)$ to plug in (and in this case we'll only need the second 0, the y-value)

$y > -3 \rightarrow 0 > -3$? Yep. We'll shade *above* the line. Done!

Answer: (see graph above)

2. $y < 3x + 1$

3. $y \geq -2x - 4$

4. $x \leq 5$

5. $2x + 3y > 6$

(Answers on pp. 405–6)

Takeaway Tips

 Equations for lines describe a *relationship* between the *x*- and *y*-values that is shared by all the (x, y) pairs on the line. Linear inequalities describe a *relationship* between the *x*- and *y*-values that is shared by all the (x, y) pairs in a region (above or below a line).

 Slope-intercept form is $y = mx + b$, where *m* is the slope and *b* is the *y*-intercept. Any equation for a (non-vertical) line can be written this way by solving for *y*.

To find the *x*-intercept, just plug **0** for *y* into the equation and solve for *x*.

To verify that a point is on a line, just plug the (x, y) values into the equation for the line and see if you get a true statement!

To graph linear inequalities, first graph the line you get from using an equals sign, use the correct kind of line—dotted or solid—and then plug a point that is *not* on the line into the original inequality to see which side to shade.

Speaking of Relationships . . .

On a serious note, from my sister, Crystal:

Did you know that there are women all over this country who live in fear of domestic violence from their husbands, but they do not try to break free?

This might blow your mind, but this tragedy makes more sense when you realize what many of them have in common: *complete financial dependence on their husbands.* Many of these women are too afraid to leave because they think they could not support themselves. Through my law practice, I've had the opportunity to represent *pro bono* (for free) several such women who have been victims of domestic violence. Not one of the women I represented had a college education, and none of them had pursued careers. They may have assumed they'd find a husband who would support them. Then, lacking job skills and feeling powerless, they believed they had no options other than to endure verbal and physical brutality, even at the risk of losing their lives.*

Focusing on your education, having a well-paying job, and being responsible about your finances (i.e., not running up a ton of credit card debt that could saddle you for years and years) can do lots of things for you. It can give you the flexibility to afford a great apartment or house as well as the fabulous clothes, shoes, and accessories that you always admired but never thought you could afford—while still maintaining a healthy bank account.

And it can also help to keep you safe.

Crystal McKellar

- - - - - - - - - -

* More than one-third of murdered women in this country are murdered due to domestic violence by their husbands or boyfriends, according to the Bureau of Justice Statistics Crime Data Brief, Intimate Partner Violence, 1993–2001, February 2003.

Chapter 11

Mystery Woman

Writing Equations for Lines, Including Parallel/Perpendicular Lines

$\mathcal{D}$o you like a good mystery? Well, grab your detective hat and get ready, because we've got some mysteries to solve. In the last chapter, we were always *given* the equation for a line. This time around, we'll get clues and have to figure them out for ourselves.

Solving a Mystery: Finding Equations for Lines

We know the equation for any (non-vertical) line can be written as $y = mx + b$, where m and b are numbers, right? So our goal is to use *clues* to fill in the m and b values. For example, if someone says, "A line has a slope of –5 and a y-intercept of 8; find an equation for this line," then all we have to do is plug in –5 for m, and 8 for b, and we get **$y = -5x + 8$**. Ta-da! Of course, most of the time, the clues aren't quite that easy.

Instead, imagine the clues are that the line has a slope of 3 and it passes through the point (–2, 4). What do we do? Taking it one step at a time, we know that $m = 3$, so our mystery equation will have to look something like this:

$$y = 3x + b$$

The equation is not quite as mysterious anymore; all that's left to do is to find b, and we'll be done! Hmm, how can we use the other clue, the point (–2, 4), to find b? Well, remember from the last chapter that whatever a line's equation is, it describes a *relationship* that is true for all the (x, y) pairs that it passes through.

In this case, we know that the point (–2, 4) satisfies the *relationship* described by $y = 3x + b$. (Think about that for a second.) So, if we plug the values **x = –2** and **y = 4** into the semi-mysterious equation $y = 3x + b$, we can find out what b would *have* to equal, in order for the equation to be a true statement and for everything to be satisfied. Plugging in the point (–2, 4), we get:

$$y = 3x + b$$

$$\rightarrow \mathbf{4} = 3(\mathbf{-2}) + b$$

$$\rightarrow 4 = -6 + b$$

$$\rightarrow \mathbf{10} = \boldsymbol{b}$$

We've found the value of b in our mysterious equation, so we can plug it in:

$$y = mx + b$$

$$\rightarrow \boldsymbol{y = 3x + 10}$$

Now that we know the values of m and b, the equation is no longer a mystery at all: Yes, we have sleuthed ourselves a line.

Step By Step

Finding the equation for a line when given the slope and a point:

Step 1. Write down the mystery equation $y = mx + b$, filling in the slope for m.

Step 2. Time to find b. Take the point you've been given, (x, y), plug those values into this semi-mysterious equation, and solve for b.

Step 3. Now that you have m and b, you can write the $y = mx + b$ form of the line.

And... Action! Step By Step In Action

A line has a slope of $-\frac{1}{5}$, and it passes through the origin. What's its equation?

Step 1. One step at a time: We know the slope, so we can write $y = -\frac{1}{5}x + b$, right?

Step 2. Let's find the value of b that makes (0, 0) satisfy our semi-mysterious equation. Plugging in $x = 0$ and $y = 0$, we get $0 = -\frac{1}{5}(0) + b \rightarrow 0 = b$. So the y-intercept is **0**, which actually makes sense because the line does pass through the origin, after all.

Step 3. Since $m = -\frac{1}{5}$ and $b = 0$, our equation is $y = -\frac{1}{5}x + 0 \rightarrow y = -\frac{1}{5}x$.

Answer: $y = -\frac{1}{5}x$

Doing the Math

Find the equation of a line from the following clues. I'll do the first one for you.

1. slope = –1, passing through (1, 2)

<u>Working out the solution</u>: Since $m = -1$, our equation will look like this: $y = -1x + b \rightarrow y = -x + b$. To find b, we'll plug in the point (1, 2) to get $y = -x + b \rightarrow 2 = -(1) + b \rightarrow 2 = -1 + b \rightarrow 3 = b$. And now we just fill in $b = 3$, to get $y = -x + 3$. Done!

Answer: $y = -x + 3$

2. slope = 2, passing through (3, 3)

3. slope = 3, passing through (1, 2)

4. slope = –1, passing through (0, 0)

5. slope = $\frac{1}{4}$, passing through (8, 6)

(Answers on p. 406)

What's the Deal?

Point-Slope Formula

Another way to find the equation of a line from its slope and a point is to use the "point-slope" formula, but I gotta tell ya, it's not any easier or faster than what we've been doing. Given a specific point (x_1, y_1) and a slope m, we can plug them into the formula:

$$(y - y_1) = m(x - x_1)$$

...and solve for y. This is very reminiscent of the formula for slope, which we'll see on p. 158, rewritten as $m = \dfrac{y - y_1}{x - x_1}$. Check out danicamckellar.com/hotx for more. But unless you *have* to learn it, trust me—your time is better spent with $y = mx + b$.

More Kinds of Clues:
Parallel and Perpendicular Lines

If two lines are parallel to each other, it means they have the same slope. For example:

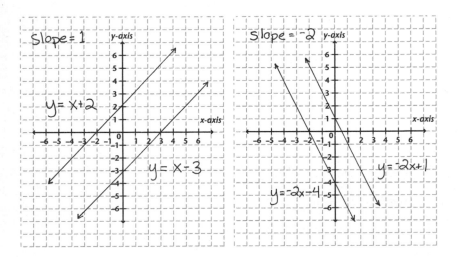

On the other hand, if two lines are perpendicular to each other, meaning they cross each other at a 90-degree angle (in other words, a right angle), they look like this:

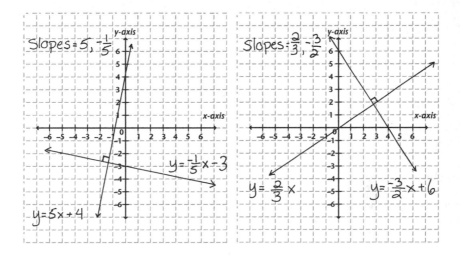

Check it out: The slopes of perpendicular lines are **negative reciprocals** of each other!

What's It Called?

The **negative reciprocal** of a number is the opposite of its reciprocal. For example, since the reciprocal of $\frac{2}{3}$ is $\frac{3}{2}$, the *negative reciprocal* of $\frac{2}{3}$ is $-\frac{3}{2}$. Just take your number in fraction form, flip it upside down, and change its sign.* For example, the negative reciprocal of 4 is $-\frac{1}{4}$, and the negative reciprocal of $-\frac{1}{3}$ is **3**.

When the slopes of two lines are *negative reciprocals* of each other, this means the lines will be **perpendicular**; in other words, they will cross each other at a right angle.

· · · · · · · · · ·

* To review reciprocals, see p. 43.

QUICK NOTE The negative reciprocal of a negative reciprocal gives you the original number again. For example, the negative reciprocal of 5 is $-\frac{1}{5}$, and the negative reciprocal of $-\frac{1}{5}$ is $\frac{5}{1} = 5$.

At some point, your teacher will want you to find the equation of a line that is parallel or perpendicular to some other line and that also passes through a particular point. This means we're essentially given the slope and a point, and we need to find the line's equation. No problem!

Step By Step

Finding the equation of a line passing through a point, and parallel or perpendicular to another line:

Step 1. Rewrite the equation they've given into slope-intercept form, if it isn't already: $y = mx + b$. (How? Just solve for y!) Now we have our slope, the m, of the existing line.

Step 2. If we're supposed to find a line *parallel* to this one, we'll keep the same slope: the same m. If we're supposed to find a line *perpendicular* to this one, we need to find the negative reciprocal of this m, and then we'll have the slope for our new line.

Step 3. Now we have our slope and a point, and we can just follow the Step By Step on p. 148 to finish. (Just stick in the slope—our new m—to create a semi-mysterious equation; then plug in the point where we see x and y to solve for b!)

And... Action! Step By Step In Action

Find the equation for the line parallel to $8x - 4y = 13$, passing through the point $(1, 1)$.

Step 1. First, we have to figure out the slope, so let's write the equation in slope-intercept form by solving for y. Subtracting $8x$ from both sides, we get $-4y = -8x + 13$. Multiplying both entire sides by $-\frac{1}{4}$, we get

$-\dfrac{1}{4}(-4y) = -\dfrac{1}{4}(-8x + 13) \quad \rightarrow \quad y = \dfrac{8}{4}x - \dfrac{13}{4} \quad \rightarrow \quad y = 2x - \dfrac{13}{4}$. So the slope of the existing line is **2**. (Notice that we don't care about the $-\dfrac{13}{4}$; the only thing we wanted from this line was its slope.)

Step 2. We want a line that is *parallel* to this one, so our new slope will also be **2**. Great! With this clue, we know our equation will look like this: $y = 2x + b$.

Step 3. Our point is **(1, 1)**, so let's plug $y = 1$ and $x = 1$ into our semi-mysterious equation, $y = 2x + b$, and discover b's value:

$$\mathbf{1} = 2(\mathbf{1}) + b \quad \rightarrow \quad 1 = 2 + b \quad \rightarrow \quad -1 = b$$

So, **$b = -1$**. We know $m = 2$ and $b = -1$, so our new equation is **$y = 2x - 1$**. Done!

Answer: $y = 2x - 1$

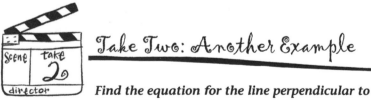

QUICK NOTE   To find the negative reciprocal of a decimal, first change it into a fraction. So what's the negative reciprocal of 0.5? Just rewrite it as a fraction and use the copycat $\dfrac{10}{10}$, so

$0.5 = \dfrac{0.5}{1} = \dfrac{0.5 \times 10}{1 \times 10} = \dfrac{5}{10} = \dfrac{1}{2}$. Now you're ready to flip it upside down and change its sign: -2. So the negative reciprocal of 0.5 is -2.

Take Two: Another Example

Find the equation for the line perpendicular to
$-y = 1.25x - 3$, passing through (5, 4).

Step 1. This line is almost in slope-intercept form. But look, the y has a negative sign in front of it. Yikes! If we didn't notice that, we might have thought the slope was 1.25. Let's multiply both entire sides by -1 (remember to distribute!) and get $y = -1.25x + 3$. So, the slope of the *existing* line is **-1.25**.

Step 2. We want to find a line *perpendicular* to this one, so our new line's slope will be the negative reciprocal of -1.25. To find this, we need to first write -1.25 as a fraction: $\frac{-1.25}{1}$ (multiply top and bottom by 100) $= \frac{-125}{100}$ (reducing) $= \frac{-5}{4}$. Ah, much better! Remember, this is still the slope of the *existing* line. We are ready to flip it and change its sign: $\frac{4}{5}$. Now we have the slope of our new, perpendicular line.

Step 3. With a slope of $\frac{4}{5}$, we know the mystery equation will look like this: $y = \frac{4}{5}x + b$. Let's use the point we were given, (5, 4), to find b. Plugging in $x = \mathbf{5}$ and $y = \mathbf{4}$:

$$y = \frac{4}{5}x + b \quad \rightarrow \quad 4 = \frac{4}{5}(5) + b \quad \rightarrow \quad 4 = 4 + b \quad \rightarrow \quad 0 = b$$

So, $\boldsymbol{b = 0}$.

Step 4. No more mystery! Because $b = 0$, $\boldsymbol{y = \frac{4}{5}x + 0}$ is the equation of our new line, which is perpendicular to $-y = 1.25x - 3$ and passes through the point (5, 4).

Answer: $\boldsymbol{y = \frac{4}{5}x}$

Horizontal and Vertical Lines

Remember our self-centered horizontal and vertical lines from p. 139?

Selfish or not, how would we find the line perpendicular to the (horizontal) line $\mathbf{y = 3}$, passing through the point (4, 6)? I mean, we usually start by finding the slope, but what's the negative reciprocal of 0? As you well know, $\frac{1}{0}$ is undefined in math, and so is $\frac{-1}{0}$, for that matter. But this is perfect because as we've seen, that's what we call the slope of vertical lines: *undefined!* And these vertical lines take the form $x = $ number, where the x-value is always that number. So if we want a vertical line to pass through the point (4, 6), we just pick the line $\boldsymbol{x = 4}$. See the graph on p. 155 for another example, and remember:

***Zero* and *undefined* slopes are indeed perpendicular to each other.**

Take Three: Yet Another Example

Find the equation of the line perpendicular to the line x = –2, passing through (2, 3).

This is a weird one—let's graph it so we can see what's going on; it might be easier than trying to follow the steps. For the line $x = -2$, the slope is undefined, and we know any line perpendicular to this must have a slope of zero. This means our line will look like y = number. And y will always be that number on the line, so in order to pass through the point (2, 3), that "number" must be 3: The answer is $y = 3$. Weird, right?

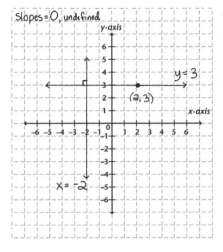

Answer: $y = 3$

 Doing the Math

Find the equation for the line described, and then graph both lines: the one given and the new line you've found. I'll do the first one for you.

1. Perpendicular to the line $y = 5x - 1$, passing through the origin

<u>Working out the solution</u>: This is already in slope–intercept form, and its slope is 5. As we've seen, the negative reciprocal of 5 is $-\frac{1}{5}$. We saw on pp. 148–49 that the equation for a line with slope $-\frac{1}{5}$ passing through (0, 0) is $y = -\frac{1}{5}x$. To graph this, let's pick easy x–values like 0, –5, and 5. Plugging these x–values into $y = -\frac{1}{5}x$, one at a time, we discover the points

(0, 0), (−5, 1), and (5, −1), which we can plot and draw a line through. For the original line, y = 5x − 1, let's pick the x–values 0, 1, −1. Plugging them into y = 5x − 1 one at a time gives the points (0, −1), (1, 4), and (−1, −6) for us to plot and draw a line through. We notice that yep, these lines look perpendicular. Done!

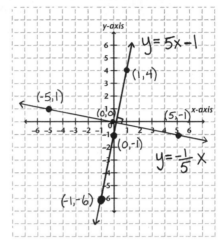

Answer: $y = -\dfrac{1}{5}x$, and see graph to the right

2. Parallel to the line $y = 3x + 3$ and passing through (0, 0)

3. Perpendicular to the line $y = 3x + 3$ and passing through (0, 0)

4. Perpendicular to the line $y = 6$ and passing through (2, 4) *(Hint: See the gray box on p. 154.)*

5. Parallel to the line $2x + 3y = 5$ and passing through (3, 1)

6. Perpendicular to the line $2x + 3y = 5$ and passing through (3, 1)

7. Perpendicular to the line $y = -0.25x + 1$ and passing through (1, 1)

(Answers on pp. 406–7)

"*I like math because I feel so clever when I figure out the answer.*"
Maddie, 13

Finding Slope from Two Points . . .
and Playing Favorites

Sometimes we won't be given the slope or even a direct clue about the slope (like that it's parallel or perpendicular to another line). Sometimes we'll only get two *points* on the line. Luckily, these are clues enough for sharp detectives like us.

Say we're given two points, like (–1, 0) and (3, 2). I'm going to show you how to find the *slope* of the line connecting them. (Slope is always a good place to start.) Slope = $\frac{rise}{run}$, and you can see on the graph that when moving from (–1, 0) to (3, 2), the *rise* is **2** and the *run* is **4**. This means the slope = $\frac{rise}{run} = \frac{2}{4} = \frac{1}{2}$. Remember, the *rise* is the change in *y*-values, and it always goes on the <u>top</u> of the slope fraction. I've always thought that a *y* on top of an *x* looks like it's lounging on a folding lawn chair, perhaps sipping lemonade.

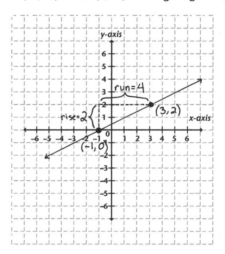

Anyway, we don't even need a graph to find the $\frac{rise}{run}$. If we know two points, like (–1, 0) and (3, 2), we can actually find the *rise* by finding the *difference* between the *y*-values—by subtracting them! And we find the *run* by subtracting the *x*-values.

I know it's not nice to play favorites, but that's exactly what we're going to do. We have to pick a favorite point: (–1, 0) or (3, 2)? Hmm, let's pick (3, 2). This means it's our point #1, and it gets to go in the **front** of the subtraction both times:

$$\text{For } \textbf{(3, 2)} \text{ and } (–1, 0): \quad \frac{rise}{run} = \frac{change\ in\ y}{change\ in\ x} = \frac{\textbf{2} - 0}{\textbf{3} - (–1)} = \frac{2}{4} = \frac{1}{2}$$

And we got the same answer as if we'd "counted" rise and run on a graph.

In fact, when we're given any two points, we could label them (x_1, y_1) and (x_2, y_2). Then <u>the slope of the line connecting these two points</u> would be:

$$\textbf{Slope} = \frac{\textbf{\textit{y}}_1 - \textbf{\textit{y}}_2}{\textbf{\textit{x}}_1 - \textbf{\textit{x}}_2}$$

Don't let those little subscripts scare you; it just shows us that our favorite point is up front with its "1" subscripts to indicate how special it is to us. And the other point gets "2" subscripts because it's, um, "point number 2." (Don't take that the wrong way.)

Let me emphasize: It doesn't matter which point is our favorite, but we *must* pick one. Once we do, it's very helpful to label the four numbers with x_1, y_1, x_2, and y_2. For example, in the previous problem, where we were given the points (3, 2) and (–1, 0), we could have picked **(–1, 0)** as our favorite point. In this case, $x_1 = -1$, $y_1 = 0$, $x_2 = 3$, and $y_2 = 2$.

For (–1, 0) and (3, 2): $\dfrac{rise}{run} = \dfrac{\textbf{\textit{y}}_1 - \textbf{\textit{y}}_2}{\textbf{\textit{x}}_1 - \textbf{\textit{x}}_2} = \dfrac{\textbf{0} - 2}{-\textbf{1} - 3} = \dfrac{-2}{-4} = \dfrac{2}{4} = \dfrac{1}{2}$

And yep, we got the same answer!*

𝒲𝒶𝓉𝒸𝒽 𝒪𝓊𝓉!

The importance of playing favorites:

For the points (3, 2) and (–1, 0), the slope fractions below would be WRONG, because even though the *y*-values are indeed on top, neither point is consistently at the front of the subtraction. In other words, there's no favorite point!

Both $\dfrac{2 - 0}{-1 - 3}$ and $\dfrac{0 - 2}{3 - (-1)}$ are <u>wrong</u> for the pair (3, 2) and (–1, 0).

So boldly declare to yourself (but um, not necessarily out loud) which <u>point</u> is your favorite, put it at the <u>front</u> of the subtractions, and you'll do great.

. . . .

* Many textbooks reverse the order of the subscripts: slope = $\dfrac{y_2 - y_1}{x_2 - x_1}$. It just goes to show that it doesn't matter which point we pick as our favorite. Because of negative signs canceling, both formulas will give the same answer! I like the 1's in front, don't you? (But do check to see if your teacher has a preference.)

QUICK NOTE   I recommend rewriting the slope formula every time you use it: $m = \dfrac{y_1 - y_2}{x_1 - x_2}$. Here's a summary for how to remember the order of the variables:
1. The y-values must be on top—remember the lawn chair.
2. Our favorite point, whose coordinates we label x_1 and y_1, goes at the **front** of both subtractions.

Step By Step

Finding the equation for a line when given two points:

Step 1. Pick a favorite point. That will be (x_1, y_1), and it will go in front of the subtractions in the slope fraction. Then label the four numbers with x_1, y_1, x_2, and y_2, and write the formula: $m = \dfrac{y_1 - y_2}{x_1 - x_2}$. Plug in the numbers, simplify the fraction, and that's the slope, m.

Step 2. Pick one of the points. Now, because we have the slope and a point, we can use these clues to write our semi-mysterious equation, and then to find b and our equation for the line! (See the Step By Step on p. 148 to review.)

And... Action! Step By Step In Action

Find the equation of the line connecting the points (–90, –240), (310, –200).

I'm glad we don't have to plot these—we'd need pretty big graph paper. Let's not worry about how big the numbers are. We'll just do the steps and see what happens.

Step 1. Let's pick **(310, –200)** to be our favorite point, so $x_1 = 310$, $y_1 = -200$, $x_2 = -90$, and $y_2 = -240$. So we get $\dfrac{y_1 - y_2}{x_1 - x_2} = \dfrac{-200 - (-240)}{310 - (-90)} = \dfrac{-200 + 240}{310 + 90} = \dfrac{40}{400} = \dfrac{1}{10}$. Okay, so we have the slope: $m = \dfrac{1}{10}$. Good progress.

Step 2. Now our equation is only semi-mysterious: $y = \frac{1}{10}x + b$. Time to find b. We can choose either point to plug in. Let's pick, I don't know, how about the point $(-90, -240)$. We'll use $y = -240$ and $x = -90$. Let's plug 'em in!

$$y = \frac{1}{10}x + b$$

$$\rightarrow \quad -240 = \left(\frac{1}{10}\right)(-90) + b$$

$$\rightarrow \quad -240 = -9 + b$$

$$\rightarrow \quad -231 = b$$

So the y-intercept is $b = -231$. Wow, this line crosses the y-axis *way* down there!

Step 3. Now our equation is $y = \frac{1}{10}x - 231$.

Answer: $y = \frac{1}{10}x - 231$

Doing the Math

Find the *slope* of the line connecting these two points and then the *equation* for the line in slope-intercept form. I'll do the first one for you.

1. $(3, 0)$, $(0, -4)$

<u>Working out the solution</u>: Let's write the slope fraction. We'll pick **(3, 0)** to go at the front of the subtractions. Let's label: $x_1 = 3$, $y_1 = 0$, $x_2 = 0$, $y_2 = -4$, and we'll get

$m = \dfrac{y_1 - y_2}{x_1 - x_2} = \dfrac{0 - (-4)}{3 - 0} = \dfrac{0 + 4}{3} = \dfrac{4}{3}$. So $m = \dfrac{4}{3}$,

and our equation will look like $y = \frac{4}{3}x + b$. To find b, let's pick $(3, 0)$, and plug it in: $0 = \left(\frac{4}{3}\right)(3) + b \rightarrow 0 = 4 + b$

$\rightarrow$ $b = -4.$* So our equation is $y = \frac{4}{3}x - 4$. Done!

Answer: $m = \frac{4}{3}$; $y = \frac{4}{3}x - 4$

2. (3, 5), (2, 4)

3. (2, 0), (0, 1)

4. (0, 5), (3, 7)

5. (0, 5), (2, –2)

(Answers on p. 407)

Takeaway Tips

When we're given clues, we can find the equation of a line in slope-intercept form, $y = mx + b$, by finding and filling in the values for m and b.

If we know the slope (m) and a point, we can plug these into $y = mx + b$ and solve for b. Then once we have m and b, we can write our equation.

When two lines are parallel, they have the same slope. When two lines are perpendicular, their slopes are negative reciprocals of each other.

If we are given two points, we can use them to find the slope of the line connecting them. Notice that the y's are on top, and our favorite point is at the **front** of both subtractions: $m = \frac{y_1 - y_2}{x_1 - x_2}$.

. . . .

* This makes sense; we had the point (0, –4), which we know is on the y-axis, after all!

TESTIMONIAL

Courtney Parry
(Chicago, IL)
<u>Before</u>: Low self-esteem in math
<u>Today</u>: Valedictorian as a math and
economics major!

In junior high and high school, I was convinced I was no good at math. From the outside, people thought I was a "natural" at math because my grades were always A's and B's...but they didn't see the truth: I could never shake my math anxiety, and my confidence was terribly low. For me, learning math was like struggling in the ocean, being hit with wave after wave whenever a new topic was introduced, almost drowning every time. It didn't help that starting in grade school, my friends would tell stories about how their siblings did horribly in algebra. I heard so many stories about how "hard" math was that I never thought it could be enjoyable or that I could be good at it. I didn't realize that it was all in my head—I had built up math as such a scary subject that I ruined my confidence!

> "When it came time for algebra, I was prepared for the worst."

When it came time for algebra, I was prepared for the *worst*. I studied extra hard and did math problems that weren't even assigned to us. The funny thing is, all this extra work really paid off. Math class started feeling like a "safer" place, and my self-esteem in math began to improve.

It still took me a long time to realize that I was good at math and to value (and embrace!) this ability. This happened for me in college, when I began to take higher-level math classes and I realized, "Hey, I can do this!" I found that the more I *relaxed* my mind, the more I began to truly enjoy it. It felt so good to overcome my fears in math; it's one of the things I'm most proud of.

Believe it or not, I recently graduated from Saint Mary's College Notre Dame with a double major in mathematics and economics—as valedictorian! After graduation, I started as an intern at an economic consulting firm in Chicago. It's really fun! Economists use math formulas to try to predict the future from what they know about the past. In algebra, the slope of a line can help you figure out where the line is headed—are things getting better or worse? In the same way, algebra can be used to look at the future and try to make a prediction about anything from how expensive houses will be next month to what the population will be in the year 2050.

I'm about to start my master's degree, and my thesis will be on Nanotechnology and Solar Energy. Math has given me the right skills to explore new, clean-energy technologies that could really help the environment and perhaps even transform the way we access energy that can be used to power everything from our cars and houses to our phones and iPods! Every day, I'm reminded how much my attitude about math has changed and how much my confidence in math has grown as a result.

Chapter 12

Little Miss People Pleaser

Solving and Graphing Systems of Linear Equations

You know the type. She's syrupy sweet and tells everyone what they want to hear . . . whether she really means it or not. Some girls just try too hard to be liked. And the truth is, you can't please everyone all of the time. That's just life. So you might as well be true to yourself, know what I mean?

Systems of equations, however, love people-pleasers because they aren't always easy to satisfy, so they need that kind of attention.

In order to satisfy a nice, normal equation like $3x + 1 = 7$, we know x must equal 2, right? That's the value of x that makes this a true statement, and everyone's happy and satisfied. But what if we had to make two equations happy, with *two* variables?

$$y = 7 - 3x$$

$$5x - 2y = 8$$

Yes, we're going to be little miss people pleasers ourselves and find the special x & y combo that pleases *both* equations. A pair of linear equations like this is also called a **system of linear equations**. Later on, we'll see how we can actually graph these equations as lines, and that the intersection point of the two lines is indeed the (x, y) combo that satisfies both equations. Pretty crazy, huh?

But graphing isn't a reliable way of *finding* this special x & y pair. Even with graph paper and good handwriting, some intersection points include fractions or decimals. Yikes! Sounds like we need a better way of finding this x & y combo. In fact, here are two methods, but the goal is always the same: to boil things down to *one* equation with *one* variable . . . which you're an ace at solving.

Substitution Method

Check it out: In the equations below, we can get rid of the y variable by using the first equation and plugging in $(7 - 3x)$ wherever we see y in the second equation. Poof! Then y is gone. So, because we are given:

$$y = \mathbf{7 - 3x}$$

$$5x - 2y = 8$$

... we can substitute $(\mathbf{7 - 3x})$ wherever we see y in the second equation:

$$5x - 2y = 8$$

$$\downarrow$$

$$5x - 2(\mathbf{7 - 3x}) = 8$$

... and then we just distribute the negative 2 and solve for x by adding 14 to both sides and combining like terms.

$$\rightarrow 5x - 14 + 6x = 8$$

$$\rightarrow 11x = 22 \quad \rightarrow \quad \mathbf{x = 2}$$

Ta-da! So now we've found the x value of our special x & y combo; we're halfway done. To find the y-value, we can now plug $x = 2$ into either of the original two equations. Let's use the first equation:

$$y = 7 - 3x$$

$$\rightarrow y = 7 - 3(\mathbf{2})$$

$$\rightarrow y = 7 - 6 \quad \rightarrow \quad \mathbf{y = 1}$$

And voilà! Our magic combo is $\mathbf{x = 2}$ and $\mathbf{y = 1}$. These values will make *both* equations true. We just saw how they make the first equation true. We should plug our answer into the second original equation just to double check: $5x - 2y = 8 \quad \rightarrow \quad 5(2) - 2(1) \overset{?}{=} 8 \quad \rightarrow \quad 10 - 2 \overset{?}{=} 8$ Yep! Everyone's satisfied, and we have been very good people pleasers, if I do say so myself.

We'll graph these two lines on pp. 178–79, and you'll see that they really do intersect at the point (2, 1). This makes sense, because it's a point that is found on both lines, isn't it? After all, in order to check if a point is on a line, we just plug the point into the equation for the line and see if we get a true statement,* which we just did—for both equations! It's amazing how all this stuff is connected.

. . . .

* See p. 135 to practice verifying if a point is on a line.

By the way, if the first equation had been $y + 3x = 7$ instead of $y = 7 - 3x$, we could have just solved for y, and then proceeded with the substitution.

Step By Step

The Substitution method for solving a system of linear equations:

Step 1. Make the equations "nicer"; reduce them if they have common factors. For fractions or decimals, multiply both sides of the equation by something to get rid of them.

Step 2. In the easier equation, solve for one variable in terms of the other.

Step 3. Substitute that expression into the second equation. Now you should have an equation in one variable. Solve for that variable.

Step 4. Stick this value into either equation, and solve for the second variable.

Step 5. Check your work by plugging the x, y combo into the equation you *didn't* use in Step 4, and if you didn't make any careless mistakes along the way (happens to the best of us), then you should get a true statement. Done!

And... Action! Step By Step In Action

Find the x & y combo that makes both equations true.

$$18x + 12y = -3$$
$$-\frac{x}{5} = \frac{y}{10}$$

Step 1. Hmm, let's make these look nicer before we start. Every term in the first equation has a factor of 3, so let's multiply *both sides* by $\frac{1}{3}$, and we get:

$$18x + 12y = -3 \quad \rightarrow \quad \frac{1}{3}(18x + 12y) = \frac{1}{3}(-3) \quad \rightarrow \quad 6x + 4y = -1$$

Ah, smaller numbers; nice! For the second equation, let's multiply *both sides* by $\frac{10}{1}$ to get rid of the fractions:

$$-\frac{x}{5} = \frac{y}{10} \quad \rightarrow \quad -\frac{x}{5} \cdot \frac{\mathbf{10}}{\mathbf{1}} = \frac{y}{10} \cdot \frac{\mathbf{10}}{\mathbf{1}} \quad \rightarrow \quad -\frac{10}{5}x = \frac{10}{10}y \quad \rightarrow \quad -2x = y$$

Now our equations look like this: $6x + 4y = -1$ and $-2x = y$. They're much friendlier looking, and we kept the scales balanced the whole time.

Step 2. As it turns out, $-2x = y$ is ready for substitution. Great!

Step 3. So, we'll plug **$-2x$** in for y in the first equation:

$$6x + 4y = -1 \ \rightarrow \ 6x + 4(\mathbf{-2x}) = -1 \ \rightarrow \ 6x - 8x = -1 \ \rightarrow \ -2x = -1 \ \rightarrow \ x = \frac{1}{2}$$

Okay, we're halfway done . . .

Step 4. Let's stick this x-value into the second equation (the easier one!) and solve for y:

$$-2x = y \ \rightarrow \ -2\left(\frac{1}{2}\right) = y \ \rightarrow \ y = -1$$

So our special combo is $x = \frac{1}{2}$, $y = -1$.

Step 5. Let's plug these into the original first equation and hope to get a true statement:

$$18x + 12y = -3 \ \rightarrow \ 18\left(\frac{1}{2}\right) + 12(-1) \stackrel{?}{=} -3 \ \rightarrow \ 9 - 12 \stackrel{?}{=} -3$$
$$\rightarrow \ -3 \stackrel{?}{=} -3 \text{ Yep!}$$

Answer: $x = \frac{1}{2}$, $y = -1$

QUICK NOTE   These are two ways of saying the same thing: "Solve this system of equations" and "Find the x- & y-values that satisfy both equations."

"*I* realize now that algebra is mostly just finding numbers, and I like a good puzzle." *Jazmin, 13*

Doing the Math

Solve these systems of equations. I'll do the first one for you.

1. $3y - 2x = 2$ and $3y - x = 7$

<u>Working out the solution</u>: Both equations are reduced, and there are no fractions. Great! Hmm, the second equation has an x by itself, so let's solve for x: $3y - x = 7$ (subtracting 3y from both sides) $\rightarrow$ $-x = -3y + 7$ (multiplying both sides by –1) $\rightarrow$ $x = 3y - 7$. Okay, now we can stick in $(3y - 7)$ for x in the <u>first</u> equation and solve for y: $3y - 2x = 2$ $\rightarrow$ $3y - 2(3y - 7) = 2$ $\rightarrow$ $3y - 6y + 14 = 2$ $\rightarrow$ $-3y = -12$ $\rightarrow$ $y = 4$. Halfway done! Let's stick y = 4 into the <u>second</u> equation and solve for x, so: $3y - x = 7$ $\rightarrow$ $3(4) - x = 7$ $\rightarrow$ $12 - x = 7$ $\rightarrow$ $x = 5$. To check our answers, let's plug them into the first equation:

$3(4) - 2(5) \stackrel{?}{=} 2$ $\rightarrow$ $12 - 10 \stackrel{?}{=} 2$ Yep!

Answer: x = 5, y = 4

2. $y = 3x$ and $y - x = 4$

3. $2y + 8x = 2$ and $2y + 7x = 3$

4. $y = \frac{x}{8}$ and $2x - 4y = 3$

(Answers on p. 407)

QUICK NOTE   This substitution method works best when one of the equations has a variable that's easy to isolate.

The Adding/Subtracting Twins Method (AKA, the Elimination Method)

Let's say we need to find the x and y values that satisfy the equations below:

$$3x + 2y = 7$$

$$5x - 2y = 17$$

Hmm . . . I don't know about you, but I'm in no mood for all the fractions we'd get if we tried to use the substitution method by solving for one of the variables.

I call this one the "Adding/Subtracting Twins" method, because that's what we do: We either *add* the two equations together, or we *subtract* them from each other. And because we're clever, we do this in a way that eliminates one of the variables. (I'll explain the "twins" part in a moment.)

First, why are we even allowed to add or subtract equations? Let's say we have this little equation:

$$3a = 5$$

If somebody told us that **b = 7**, then we could safely add b to one side of the above equation and 7 to the other side, and we would still have a true statement: $3a + b = 5 + 7$. See what I mean? After all, since $b = 7$, we are keeping the scales perfectly balanced. Another way to write this would be:

$$3a = 5$$

$$+ \underline{\quad b = 7 \quad}$$

$$3a + b = 12$$

We could also *subtract* these equations from each other, and we'd end up with yet another true statement. Because b equals 7, we're just taking away something equal from both sides of the equation $3a = 5$, and the scales still remain balanced:

$$3a = 5$$

$$- \underline{\quad b = 7 \quad}$$

$$3a - b = -2$$

In the same exact way, we can add more complicated-looking equations together. I mean, what's the difference? That equals sign means the two things on either side of it are *equal*, after all.*

Let's add these two equations together, $3x + 2y = 7$ and $5x - 2y = 17$, and see what kind of true statement we get:

$$3x + 2y = 7$$
$$+ \quad \frac{5x - 2y = 17}{8x + 0 = 24}$$
$$\rightarrow \quad 8x = 24$$

And lookie there—because we added $2y$ to $-2y$, the y terms went away! What we're left with is easy to solve. Dividing both sides by 8, we see that **$x = 3$**. Then we can pick either of the first equations to plug in 3 wherever we see x, and we'll find that **$y = -1$**. And there's our x & y combo that makes both equations true.

QUICK NOTE   Notice that this worked so well because we had a set of "twins": $2y$ showed up in both equations. Well, it was $-2y$ in one of them . . . that must be the evil twin. And that's our first goal in this method—to get twins, even if we didn't start out with them.

Step By Step

Adding/Subtracting Twins (Elimination) method for solving systems of linear equations:

Step 1. Make sure you have twins! In other words, one of the variable terms in one equation needs to be the <u>same</u> or the <u>opposite</u> of one in the

· · · · · · · · · ·

* This stuff can take a moment or two to sink in. Don't worry if you have to read the section a couple of times.

other equation. Do things to both entire sides of the equation(s) until this is true, keeping the scales balanced.

Step 2. If a variable term is the <u>same</u> in both equations—identical twins—then *subtract* the equations, and watch your negative signs (I recommend scratch paper). If a variable term in one equation is the <u>opposite</u> of a variable term in the other—one is an evil twin—then *add* the two equations. Watch the twins disappear!

Step 3. Solve your new equation for its remaining one variable.

Step 4. Plug this value into either original equation, and solve for the other variable.

Step 5. Check your answer by plugging these two values into the equation you didn't use in Step 4, and make sure you get a true statement. Done!

And... Action! Step By Step In Action

Solve this system of equations:

$$5x + 7y = 15$$
$$5x - 4y = -7$$

Step 1. We have identical twins! The term *5x* shows up in both equations, so if we subtract these equations, we'll be able to eliminate the *x* terms.

Steps 2 and 3. This subtraction is tricky, so we better pay attention!

$$5x + 7y = 15$$
$$-\quad \underline{5x - 4y = -7}$$
$$0 + 11y = 22$$
$$\rightarrow \quad 11y = 22$$
$$\rightarrow \quad y = 2$$

subtraction scratch work:
$$5x - 5x = 0$$
$$7y - (-4y) = 11y$$
$$15 - (-7) = 22$$

Step 4. Next, let's plug **y = 2** into the first equation, and we get:

$$5x + 7y = 15 \rightarrow 5x + 7(2) = 15 \rightarrow 5x + 14 = 15 \rightarrow 5x = 1 \rightarrow x = \frac{1}{5}$$

Step 5. Great! Now, to make sure we really got the x & y combo that satisfies both equations, let's plug **x = $\frac{1}{5}$** and **y = 2** into the *second* equation:

$$5x - 4y = -7 \rightarrow 5\left(\frac{1}{5}\right) - 4(2) \overset{?}{=} -7 \rightarrow 1 - 8 \overset{?}{=} -7 \rightarrow -7 = -7$$

Yep, we did!

Answer: $x = \frac{1}{5}$, $y = 2$

Watch Out!

Especially when subtracting, pay close attention to the negative signs that already exist; it can get tricky! If you don't show your work, at least write it out on scratch paper somewhere. It's just too easy to make mistakes.

$$5x + 7y = 15$$
$$-\underline{\quad 5x - 4y = -7 \quad}$$
$$0 + 3y = 8$$
$$\rightarrow 3y = 8 \rightarrow y = \frac{8}{3}$$

"Oh, I don't need any silly scratch paper..."

... NOT!

Yikes! Just use that scratch paper, and you'll be fine. (C'mon, it takes like two seconds.)

QUICK NOTE   Often, we'll have to multiply both sides of one (or both!) of the equations by something in order to get twins, evil or not. Remember, as long as we do the same thing to both entire sides of an equation, we'll keep the scales balanced.

Take Two: Another Example

Find the x & y combo that makes both equations true:

$$15x + 10y = 0$$

$$2x - 5y = 19$$

Step 1. If we multiply both sides of the second equation by 2, we'll end up with $10y$ "evil twins," which is great! So we get $2x - 5y = 19 \rightarrow$ $2(2x - 5y) = 2(19) \rightarrow$ **$4x - 10y = 38$**. We'll *add* them in order for the $10y$ twins to disappear.

Steps 2 and 3. Now we're ready to add and solve for x:

$$
\begin{aligned}
15x + 10y &= 0 \\
+\quad 4x - 10y &= 38 \\
\hline
19x + 0 &= 38
\end{aligned}
$$

$$\rightarrow\ 19x = 38 \ \rightarrow\ \underline{x = 2}^{*}$$

Step 4. Okay! Now that we know **$x = 2$**, we can plug it into either equation; let's pick the first one: $15x + 10y = 0 \ \rightarrow\ 15(2) + 10y = 0 \ \rightarrow\ 30 + 10y = 0$ $\rightarrow\ 10y = -30 \ \rightarrow\ \textbf{y = -3}$.

Step 5. To check our work, let's plug the combo **$x = 2$ & $y = -3$** into our original second equation to see if it's satisfied: $2x - 5y = 19$ $\rightarrow\ 2(2) - 5(-3) \overset{?}{=} 19 \ \rightarrow\ 4 + 15 \overset{?}{=} 19 \ \rightarrow\ 19 = 19$. Yep!

Answer: $x = 2$, $y = -3$

Difficult Twins: What's the Deal?

Sometimes we need to multiply *both* equations by something in order to produce twins. For example, with $7x + 10y = 17$ and $8x + 15y = 23$, we can create $30y$ twins by multiplying both sides of the first equation by 3 and the second equation by 2:

$$\mathbf{3}[7x + 10y] = \mathbf{3}(17) \ \rightarrow\ \mathbf{21x + 30y = 51}$$

$$\text{and}\quad \mathbf{2}[8x + 15y] = \mathbf{2}(23) \ \rightarrow\ \mathbf{16x + 30y = 46}$$

. . . .

* It wasn't hard to know that $38 \div 19 = 2$; after all, we just did $19 \times 2 = 38$.

And we'd proceed from there. Notice that 30 is the lowest common multiple of 10 and 15, so 30y is the smallest twin we can make out of these equations. And small is good. By the way, we could also create 56x twins by multiplying the first equation by 8 and the second by 7, but those numbers get really big, like 23 × 7 . . . um yeah, no thanks.

Doing the Math

Solve these systems of linear equations using twins. I'll do the first one for you.

1. $x + \frac{7}{5}y = 3$ and $10x - 8y = -14$

<u>Working out the solution</u>: That fraction might be unpleasant to deal with, so let's get rid of it by multiplying both sides by 5.

We get $5\left[x + \frac{7}{5}y\right] = (5)3$ $\rightarrow$ <u>$5x + 7y = 15$</u>. We have many choices; we could make 10x twins by multiplying this equation by 2. But hey, the second equation can actually be reduced—and smaller is usually nicer. We can factor a 2 out of every term, and we get $10x - 8y = -14$ $\rightarrow$ $2(5x - 4y) = 2(-7)$ $\rightarrow$ dividing both sides by 2 $\rightarrow$ <u>$5x - 4y = -7$</u>. Well look at that! Some 5x twins have appeared in our (underlined) equations. And in fact, we did this problem on pp. 171–72, and our answer was $x = \frac{1}{5}$, $y = 2$.

Answer: $x = \frac{1}{5}$, $y = 2$

2. $5x - 3y = 19$ and $2x + 3y = -5$

3. $3x - 5y = -36$ and $3x + 5y = 12$

4. $7x + 10y = 17$ and $8x + 15y = 23$ *(Hint: See the QUICK NOTE on p. 172.)*

5. $4x + 3y = -7$ and $3x + 2y = -6$

(Answers on p. 407)

Mountain Climbing

$\mathcal{A}$s you grow into a young woman, the bigger you dream, the more likely you are to hear, "See your goal over there? You'd have to climb that impossibly enormous mountain to get it." But unlike most people who get intimidated by hard things, *you'll* say, "What mountain? Oh, you mean *that* little thing? No problem." And do you know why? Because you're practicing your math, and let's face it, sometimes math can be hard!

Decide to put yourself in the category of people who have the stamina to do hard things, and when you set a goal, don't worry if it takes you awhile to get there. Feeling impatient and giving up early is what keeps most people from reaching their goals. Do you think a runner is born being able to run a marathon? No way! But just like you, she decided to believe in herself and to be her own best friend through thick and thin.

Knowing what you want, and having the confidence and stamina to earn it, is one of the best feelings in the world. It will bring you the ultimate happiness that so many people search for their whole lives. Remember: Success happens one step at a time.

If you haven't graphed lines in awhile, I recommend reading Chapter 10 before continuing with this chapter. It'll make a whole lot more sense if you do!

She's Never Satisfied:
Graphing Systems of Linear Equations

You just can't please some people, especially the ones who don't even know what they want. One day you bring them vanilla ice cream and they turn it away, complaining, "But we like chocolate." Fine. Then the next day you bring them chocolate and they say, "But we like vanilla!" I mean, when they're **inconsistent** like that, there's no way they could ever be satisfied! It's a good reminder for us all to make up our minds and be grateful for what we are given. Plus, it'll help you remember an upcoming definition.

What Are They Called?

For a system of two linear equations:

If a pair of equations is **inconsistent**, it means there is *no* value of x and y that will satisfy both equations. (You just can't please some people.) When we try to find a solution, we'll always get a false statement! When graphed, their two lines are parallel and will never touch each other at all. For example, the system of equations $y = 3x - 1$ and $y = 3x + 5$ is *inconsistent.** No intersection point means there's <u>no</u> (x, y) pair that satisfies both equations, see what I mean?

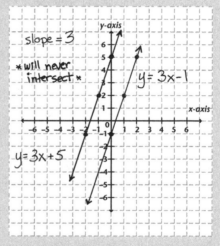

If the two equations are **consistent**, it means there is at least one x & y combo that satisfies both equations, maybe even an infinite number! Most pairs of equations you come across in your textbook are *consistent*. So that's nice and . . . consistent, isn't it?

. . . .

* These two lines have the same slope, 3, but different y-intercepts, so that's how we know they are parallel but not the same line! We found the labeled points by plugging in x = 0, 1, 2 in the line $y = 3x - 1$, and then x = -2, -1, 0 in the line $y = 3x + 5$.

You can always tell a pair of *inconsistent* equations, and not just because they're always throwing ice cream back in your face. Just write them both in $y = mx + b$ form, and if they have the same exact slope but different y-intercepts, they must be parallel, which means they never intersect.

QUICK NOTE   "A system of two equations" means the same thing as "a pair of equations."

What Are They Called?

If a system of two linear equations is *consistent* (meaning there is at least one solution—the lines are *not* parallel), then the pair will be either **independent** or **dependent**.

If a pair of equations is **independent**, there is exactly one x & y pair that satisfies them both. When graphed, their two lines cross in exactly one place, and their intersection point (x, y) is the solution. Just remember, if they're <u>in</u>dependent, then they <u>in</u>tersect at exactly one point. (See p. 179 for an example.)

If a pair of equations is **dependent**, it means that an infinite number of x & y pairs will satisfy both equations. Why? Because the equations can be rewritten to look identical. They are the *same* equation in disguise! When graphed, they are the same line, smack on top of each other, so you can't even tell there are two of them. Sounds pretty co-*dependent*, if you ask me. And in fact, every point on that single line (there are infinitely many of them) represents an (x, y) combo that will satisfy "both" equations. For example, the equations $x + y = 3$ and $2x + 2y = 6$ are the same line (just solve for y; for both, you'll get $y = -x + 3$), so they are *dependent*. See p. 181 for an example.

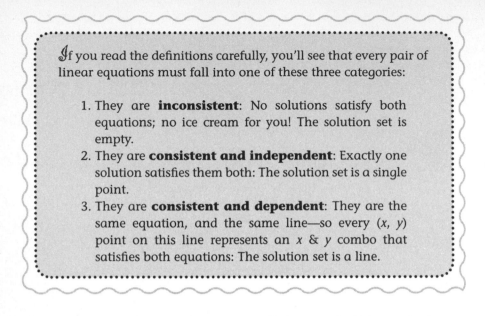

If you read the definitions carefully, you'll see that every pair of linear equations must fall into one of these three categories:

1. They are **inconsistent**: No solutions satisfy both equations; no ice cream for you! The solution set is empty.
2. They are **consistent and independent**: Exactly one solution satisfies them both: The solution set is a single point.
3. They are **consistent and dependent**: They are the same equation, and the same line—so every (x, y) point on this line represents an x & y combo that satisfies both equations: The solution set is a line.

Now that we've straightened out all this vocab, let's graph these systems of equations we've been classifying. Less talk, more action!

Step By Step

Finding the intersection of two lines and graphing the result:

Step 1. Solve the system of equations, either with the substitution method or the elimination twins method.

Step 2. If you get a true statement, then the system is dependent—they are the same line. If you get a false statement, then the system has no solution (it's inconsistent). If you get an x & y combo, like we have so far in this chapter, then that (x, y) pair will be the intersection point of the lines.

Step 3. Graph each equation as a line by finding points and plotting them, and label their intersection point, if any.

And... Action! Step By Step In Action

Solve and graph the system: $5x - 2y = 8$ and $y = 7 - 3x$.

Steps 1 and 2. To find the intersection point of these two lines, we need to find the x & y combo that satisfies both equations. On p. 165, we found that **x = 2, y = 1** is the solution that satisfies both equations, so that means the intersection point of the two lines is **(2, 1)**. Showing off a little math vocab, we can say that these two equations are *consistent* and *independent*, can't we?

Step 3. Time to graph! First, let's graph $5x - 2y = 8$. I always like to rewrite my equations into $y = mx + b$ form* before graphing them. Isolating y, we get $5x - 2y = 8 \rightarrow 2y = 5x - 8 \rightarrow$ (dividing both sides by 2) $\rightarrow y = \frac{5}{2}x - 4$. So it has a positive slope, and because –4 is the y-intercept, the line crosses the y-axis at the point **(0, –4)**. I like it when fractions go away, so let's use $x = 4$ to find another point: $y = \frac{5}{2}x - 4 \rightarrow y = \frac{5}{2}(4) - 4$ $\rightarrow y = 10 - 4 \rightarrow \mathbf{y = 6}$. Great, a second point to plot is **(4, 6)**. We already know it passes through the point **(2, 1)**, so now we plot all three points and draw a line through them (see below).

Time to graph the second equation: $y = 7 - 3x$. Again, let's put it into $y = mx + b$ form: $y = -3x + 7$. This line crosses the y-axis at 7, so one point on this line is **(0, 7)**. We also know this line passes through **(2, 1)**, and to find a third point to plot, let's use $x = 1$. We get $y = -3(\mathbf{1}) + 7 \rightarrow y = -3 + 7$ $\rightarrow \mathbf{y = 4}$. So another point is **(1, 4)**. We're ready to plot and graph the second line, and watch how they so beautifully intersect where they are supposed to: at the x & y pair that satisfies both equations.

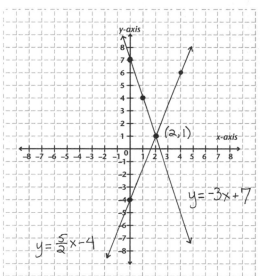

Answer: They intersect at (2, 1). See graph to the right.

Take Two: Another Example

Solve and graph this system of equations:
$y = 3x + 5$ and $6x - 2y = 2$.

Steps 1 and 2. Looking at the first equation, let's use substitution to plug in $(3x + 5)$ wherever we see y in the second equation: $6x - 2y = 2$ → $6x - 2(\mathbf{3x + 5}) \stackrel{?}{=} 2$ → $6x - 6x - 10 \stackrel{?}{=} 2$ → $-10 \stackrel{?}{=} 2$. Huh? We got a false statement! This means there is no solution, so the lines are parallel. In other words, this system is *inconsistent*. Nope, can't please some people.

Step 3. The second equation, $6x - 2y = 2$, can be rewritten as $y = 3x - 1$, and in fact, we graphed these parallel lines on p. 176.

Answer: No solution; inconsistent system (see graph on p. 176)

Shortcut Alert

If you are asked to classify a pair of equations as consistent/inconsistent, and/or independent/dependent, the fastest way is to put both equations into $y = mx + b$ form. See, if their slopes (the m's) are *different*, then just think about it: The two lines must cross each other somewhere, right? And this means the pair of equations is *consistent* and *independent*. (See pp. 177–78 to review these definitions.) But if the two slopes are the *same*, then the lines are either parallel (*inconsistent*) or they are actually the same line (*consistent and dependent*). And it should be pretty easy to tell if they are the same line: Their $y = mx + b$ forms will be identical!

Doing the Math

a. Are the equations inconsistent, consistent & independent, or consistent & dependent?

b. State which (x, y) points are solutions, if any.

c. Graph the equations as lines, and if there is an intersection point, label it. I'll do the first one for you.

1. $y = -3x + 7$ and $6x + 2y = 14$

<u>Working out the solution</u>: The fastest way to answer the first part is to put the equations in $y = mx + b$ form. The first one's done for us, so let's rewrite the second one: $6x + 2y = 14 \rightarrow$ (subtracting $6x$ from both sides) $\rightarrow 2y = -6x + 14$ (dividing both sides by 2) $\rightarrow y = -3x + 7$. Hey, this is the first equation! This means they are consistent and dependent, so every single x & y pair that satisfies the equation $y = -3x + 7$ will be a solution to this "pair" of equations. And we actually graphed this as one of the lines on p. 179.*

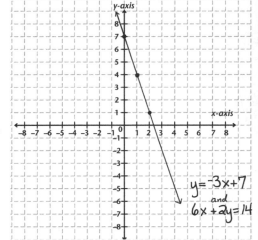

Answer: a. They are consistent & dependent; b. The solution is all (x, y) points found on the line $y = -3x + 7$; c. See graph to the right.

2. $y = -2x + 4$ and $y = 2x + 4$

3. $y = 5x + 5$ and $5y = 25x$

4. $x = \dfrac{y}{2} - 3$ and $y = 2x + 6$

5. $x + 2y = 10$ and $3x - y = 1$

6. $x + 2y = 10$ and $y = 3$ *(Remember, $m = 0$ for the line $y = 3$.)*

(Answers on pp. 407–8)

. . . .

* By the way, if we had used the substitution or elimination method on these equations to try to find their intersection point, we would have ended up with a *statement that's always true*, like 2 = 2, for example. And that would tell us the same information: They are indeed the same line.

Takeaway Tips

 For most pairs of linear equations you'll see, there will be a single *x* & *y* combo that satisfies both equations. This will be the same (*x*, *y*) point where their two lines intersect. These equations are called *consistent* and *independent*; they *intersect* at exactly one point.

 To find this special *x* & *y* combo, we can use substitution or elimination (AKA adding/subtracting twins!)

 If two equations can be written as the same exact equation, then the infinite number of (*x*, *y*) points lying on that line will satisfy both equations. These equations are *dependent*.

 If two equations have no *x* & *y* combo that satisfies them both, they are called *inconsistent* (remember the ice cream!) and their lines will never cross each other; they're parallel.

Danica's Diary/Confessional
DUMBING OURSELVES DOWN— THE SEQUEL!

We've all been tempted to dumb ourselves down around a guy so that he'll feel extra smart and special. But is it worth it? No way! Especially because there are tons of ways of making guys feel good without having to turn ourselves into ditzes! Besides, most guys don't actually want dumb girlfriends. I recently shot an episode of a TV show called *The Big Bang Theory* on CBS, and one of the cute lead actors told me he was going to have a first date that night. The next day I asked him how it was, and he said, "Eh, not great—she was kinda dumb." 'Nuff said.

I asked you to send me letters about *your* experiences, and here's what you said:

"On p. 50 of *Kiss My Math*, you told us how to make guys feel good by asking them stuff we *actually* don't know—and it worked! There's this guy, Mr. Class Clown, the ultimate popular jock. Let's say his name is 'Joe.' Many people think he's a spaz because of his antics. For some reason, I see something different in him. His jokes are funny, and his eyes, oh they're amazing. They're like liquid gold where the iris should be. Anyway, we have math class together, and when I missed class because of my vacation, I was a little freaked out . . . so I asked 'Joe' if he could teach me what I missed. Now I'm sort of a nerd, so it's not often that I don't understand something. You should have seen his face. He totally lit up and smiled. 'Well, sure!' he said. I was floating, and after he explained the entire thing to me, I totally got it. I was sooo happy, and I think he was, too! I made him feel smart, and I didn't have to dumb myself down to do it. In fact, he knows I'm smart, and I think that made it even cooler when he helped me."
Meghan, 13

"The guy I like has a learning problem, so schoolwork is really hard for him, and he's told me he wishes he could do a math page as fast as I can. I don't dumb myself down just so that he feels smarter. Instead, I show him that there are a lot of things he's really good at that I'm not, like a lot of sports! That way we both feel good at something." **Cheyenne, 14**

"I had this huge crush on a guy, but he was one of those guys who thinks girls shouldn't be as smart as guys . . . so I flirted with him a little and acted like I was dumb. I started to get his attention, but then my grades started falling. I decided it wasn't worth it—I stopped talking to the guy and focused on improving my grades. It was funny because the guy still liked me even though I was openly smart, but I realized I didn't like him anymore anyway. Take it from me—don't act dumb just to try and impress some guy! It's just not worth it." **Shawnie, 16**

Chapter 13

The Guy of Your Dreams

Motion Problems: Using $r \cdot t = d$

$\mathcal{A}$cross the field of waving grains, there he is: the guy of your dreams. He's smart, funny, totally hot, and he treats you like a queen. Standing in the fields, the sun glinting off his dazzling smile, he begins to run toward you as you run toward him. The music swells. If you start off a quarter of a mile apart and he runs at 4.75 miles per hour and you run at 4.25 miles per hour, how far and how long will you have to run before you can embrace?

Yes, these and many other types of important questions will be answered in the next few chapters, full of all your favorite types of word problems: percent increase, motion, work, mixture, and more. Aren't you just so excited? Don't worry; soon you'll boldly approach word problems without breaking a sweat.

We'll get back to the guy and the field in a moment.

Motion Problems

No, this isn't like motion sickness. These are word problems involving motion . . . also known by their formula:

$$r \cdot t = d$$

where r = rate of motion, or speed (km/hour, feet/min, etc.)

t = time (seconds, minutes, hours, days, etc.)

d = distance (km, miles, feet, inches, cm, etc.)

So for example, if you know how far away your best friend's house is (1.5 miles) and you know that it takes you 30 minutes to walk there, you can find out how fast you walk in miles/hour by plugging in the values you know and solving for the unknown.

First, since we'll want our answer to be in miles per hour, let's make sure we're dealing with miles and hours for t and d. Let's write 30 minutes as $\frac{1}{2}$ hour, and now we're ready to plug in our values for t and d:

$$r \cdot t = d$$

$$r \cdot \frac{1}{2} = 1.5$$

To isolate r, let's multiply both sides by 2:

$$r \cdot \frac{1}{2}(2) = (1.5)(2)$$

$$r = 3$$

Answer: 3 miles/hour

So you walked at a rate of 3 miles per hour to your BFF's house. Funny though, how if you talk to her the whole way over, you barely even notice you're getting exercise!

QUICK NOTE   In these motion problems, it's important that the units all match each other. So if you are working with miles/hour, make sure the time is expressed in hours and the distance is in miles. If you use *miles/hour* for the rate but *minutes* for the time, you'll get the wrong answer. Having matching units is actually why the $r \cdot t = d$ formula works—the units cancel! (For more on this, see p. 207.) Teachers try to stump us with it all the time. Just pay attention to your units and you'll be golden.

In algebra, most word problems dealing with motion will be more complicated than the one we just did. They usually involve a chart and also a system of equations, just like we learned to solve in Chapter 12. Get good at these, and you'll be an algebra star! (Also, if your pre-algebra word problem skills are rusty, I highly recommend Chapters 11 and 13 in *Kiss My Math* for a review!)

Step By Step

Solving motion word problems:

Step 1. Are the units consistent? If not, do the necessary unit conversions.*

Step 2. Identify the important information:† rates, times, and distances given, and any relationships between them. Will two *distances* be the same (as in round-trip problems)? Are the *times* the same (as in "moment in time" problems), etc.? Label the *unknowns* with the variables of your choice. Maybe draw a picture, too!

Step 3. Make an $r \cdot t = d$ chart to organize your info: The chart might depict a **round-trip** (the rows will be the "to" and "from" trips) or a **particular moment in time** (the rows will each be a different person/thing that travels). And this is the time to write the unknown variables *in terms of* the information you *do* know. Use as few variables as you can to do this. Can you now set two quantities equal to each other? Once you've used all the pertinent information, you should end up with <u>one equation only using one variable, or a system of equations in two variables</u> (like we solved in Chapter 12).

Step 4. Solve for one of the variables—whichever is easiest to solve for.

Step 5. If you haven't found the thing the problem was asking for, solve for it now.

This will make more sense when we do an example. Let's do it!

And... Action! Step By Step In Action

So, today you decided to walk from your house to the park at a nice steady pace of 3 miles per hour. On the way back, your iPod played a much faster song, energizing you, and you jogged the whole way back! In fact, you jogged 2 miles/hour faster than you walked. If the entire trip took you a total of 4 hours, how far is the park from your house?

. . . .

* To brush up on unit conversions, check out Chapter 19 in *Math Doesn't Suck*.
† Try <u>underlining</u> the important information in the problem. It makes things easier to see.

Okay, take a deep breath. Mastering these kinds of problems will give you so much confidence, soon enough you'll be like "oh yeah, bring it on."

Step 1. The units are all in hours and miles/hour, totally consistent, so no unit conversions are necessary. Fabulous!

Steps 2 and 3. In round-trip problems, the two distances will be the same—good to keep in mind. Let's make a chart! The rate for *walking* is **3** miles/hour, and the rate for *jogging* is 2 more than that, so **5** miles/hour, right? We also know that the time to the park plus the time back home will add up to equal 4 hours. Let's label **k** as the <u>time</u> it took you to *walk*. And notice that if *j* is the <u>time</u> it takes to jog, we know that $k + j = 4$ is true. And instead of writing *j* in for the jogging time, we can write *j* <u>in terms of k</u>. So $k + j = 4 \rightarrow j = 4 - k$, and we have **4 – k** for the jogging time. Does that make sense? This is a handy trick you'll use all the time to eliminate variables. First we'll fill those first two columns:

A Round-Trip Chart

Formula →	*r*	•	*t*	=	*d*
Units →	miles/hour	×	hours	=	miles
Walking There	3	×	K	=	3k
Jogging Back	5	×	4–k	=	5(4–k)

We filled in the final column using our handy formula $r \cdot t = d$, multiplying across to get *d*'s values. So, if *k* is the *time* it took you to <u>walk</u> to the park, then **3k** is one way to express the *distance* to the park. That's because $r \cdot t = d$ is always true. Make sense? And the same thing works for the jogging row: **5(4 – k)** expresses the *distance* on the way back home.

Steps 4 and 5. Here's the key to round-trip problems: We can now set the two "*d*" column values *equal* to each other, and we know we'll have a true statement. The distance from your house to the park is the same as the distance from the park to your house, so that means $3k$ and $5(4 - k)$ must be equal.

$$3k = 5(4 - k)$$

The hard part is over! Now we solve for *k*. Distributing the 5, we get:

$$3k = 5(4 - k)$$

$$3k = 20 - 5k$$

$$8k = 20$$

So, $k = \dfrac{20}{8} = \dfrac{5}{2}$, or **2.5 hours**. We were asked for the distance from your house to the park, which we know can be written as 3*k*, so $3k = 3(2.5) =$ **7.5 miles**.

Answer: The distance from your house to the park is 7.5 miles. That's quite a workout! I foresee a nap in your future.

Do your best to follow along, and <u>make sure to read these a few times</u>. They take some getting used to. And don't expect that you could have come up with these methods on your own. The more problems you see and practice, the more "intuition" you will get. That's right—math intuition comes from experience; don't let anyone tell you otherwise!

Now let's do one where the *times* are equal. I like to call these "Moment in Time" problems.

 Take Two: Another Example

Claire and the school's track star, Kim, start running around the quarter-mile track at the same time. Kim runs 2 miles/hour faster than Claire. How long will it be until Kim laps Claire?

Step 1. The units look good, but this problem is weird. It seems like it doesn't have enough information. Let's move forward anyway.

Step 2. So, what does it mean to "lap" someone? It means that Kim passes Claire somewhere on the track, so at *this moment*, Kim has run exactly 1 lap (0.25 mile, in this case) more than Claire has. It doesn't mean they've met again at the starting point; it could be anywhere along the track. The important thing is that we know *at the moment of passing*, Kim has traveled exactly 0.25 miles farther than Claire. (Let that sink in.) Check out the picture on the next page if it helps.*

· · · · · · · · · ·

* The picture suggests Kim lapped Claire after running less than a full lap. If they're fast runners, it probably took many laps—but the picture gets messy with all those footprints!

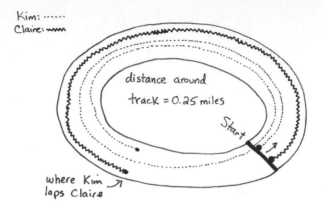

Kim: ......
Claire: ⌇⌇⌇

distance around
track = 0.25 miles

Start

where Kim
laps Claire

Let's call Claire's rate c. Then if Kim runs 2 miles/hour faster, we could call Kim's rate $c + 2$, right? We'll call the distance Claire runs D, which means we can express the distance Kim has run as $D + 0.25$, and we'll call the time of passing, t.

Step 3. Let's fill in what we know about the *moment* when Kim passes Claire. Note that we're only using one variable, t, for both times, because this chart describes a moment when they've both run for the <u>same amount of time</u>, so the times must be equal!

The <u>Moment</u> Kim Laps Claire:

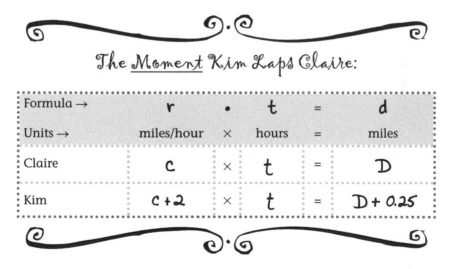

Formula →	r	•	t	=	d
Units →	miles/hour	×	hours	=	miles
Claire	c	×	t	=	D
Kim	$c + 2$	×	t	=	$D + 0.25$

Steps 4 and 5. Hmm, two equations but three unknown variables? Let's just have a little faith and see what happens. Writing out the above two $r \cdot t = d$ equations:

$$ct = D$$

$$(c + 2)t = D + 0.25$$

Using $ct = D$, we'll substitute ct for D in the second equation and then distribute:

$$(c + 2)t = \boldsymbol{D} + 0.25$$

(And, since $\boldsymbol{ct = D}$)

$$\rightarrow (c + 2)t = \boldsymbol{ct} + 0.25$$

$$\rightarrow ct + 2t = ct + 0.25$$

Almost miraculously, we can subtract ct from both sides, and the c variable completely goes away! Now we're left with $\boldsymbol{2t = 0.25}$, so dividing both sides by 2 and simplifying, we get:

$$t = \frac{0.25}{2} = \frac{0.25 \cdot \boldsymbol{100}}{2 \cdot \boldsymbol{100}} = \frac{25}{200} = \frac{1}{8}$$

Answer: In $\frac{1}{8}$ hour (7.5 minutes),* Kim will lap Claire.

We've learned that Kim's and Claire's speeds could be anything. As long as Kim runs *2 miles/hour faster* than Claire, she'll lap her in $\frac{1}{8}$ hour!

You can plug numbers into $r \cdot t = d$ if you want to see how that works: Let's say Claire runs at 4 miles/hour, and Kim runs at 6 miles/hour (which is indeed 2 miles/hour faster). How far will each have gotten in $\frac{1}{8}$ hour? Are the two distances different by exactly 0.25 mile—in other words $\frac{1}{4}$ mile? (The answer is yes. They'll have run $\frac{1}{2}$ mile and $\frac{3}{4}$ mile, respectively.) Try two other rates—5 miles/hour and 7 miles/hour. It'll work again. Pretty neat, huh!

QUICK NOTE: Currents and Windspeeds
Some problems involve a windspeed or a river's current. These are *rates* that get added to or subtracted from the *rate* of the actual person, boat, airplane, and so on.

Take Three: Yet Another Example

The wind whipping through her hair (and slowing her down), Kayla left the volleyball nets and Rollerbladed up

. . . .

* Just divide 60 minutes by 8 to get 7.5 minutes (by *hand*, missy, it's not hard).

the beach's boardwalk to the ice-cream stand. *When there is no wind, Kayla skates at 6 miles/hour. On her return trip, the wind was at her back, and it made her go faster. (It also made her hair fly into her ice cream from time to time.) Her total travel time was 36 minutes, and it took her twice as long to get there as it did to come back. What was her speed heading back to the volleyball nets?*

Step 1. First, our units need some adjusting. We've got miles per hour and minutes. Converting the 36 minutes into hours:

$$36 \text{ min} = \frac{36 \cancel{\text{ min}}}{1} \cdot \frac{1 \text{ hour}}{60 \cancel{\text{ min}}} = \frac{36}{60} \text{ hours} = \frac{3}{5} \text{ hours}$$

Steps 2 and 3. Let's call the time it takes to <u>go</u> to the ice-cream stand g and the time coming <u>back</u> b. We've been told that the time going there was twice the time coming back, so **$g = 2b$**. We also know that the total travel time equals $\frac{3}{5}$ hour. So **$g + b = \frac{3}{5}$**. We can substitute $2b$ for g and get **$2b + b = \frac{3}{5}$**, so $3b = \frac{3}{5}$, right? Solving for b, we get $b = \frac{1}{5}$ hours, which means that $g = 2b = 2\left(\frac{1}{5}\right) = \frac{2}{5}$ hours. So far, so good? Let's write those *times* in the chart. Now, about rate: We know that Kayla's Rollerblading speed *without* wind is 6 miles/hour. If we call the windspeed w (also a rate), then her speed going to the ice-cream stand is $6 - w$, because the w speed is slowing her down. On the way back, the wind makes her go faster, so her speed could be written as $6 + w$. Does that make sense? Let's put this in a chart and multiply across, because $r \cdot t = d$:

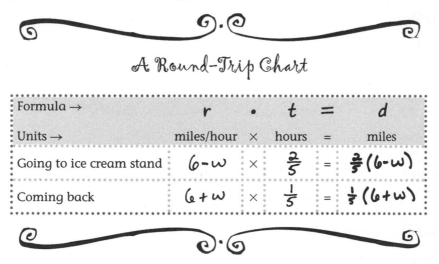

A Round-Trip Chart

Formula →	r	•	t	=	d
Units →	miles/hour	×	hours	=	miles
Going to ice cream stand	$6-w$	×	$\frac{2}{5}$	=	$\frac{2}{5}(6-w)$
Coming back	$6+w$	×	$\frac{1}{5}$	=	$\frac{1}{5}(6+w)$

Step 4. The distance to the ice-cream stand is the same as the distance coming back, so we set the two d's equal to each other. (First step? Multiply both sides by 5 to get rid of those fractions.)

$$\frac{2}{5}(6 - w) = \frac{1}{5}(6 + w)$$

$$\rightarrow \quad 5 \cdot \frac{2}{5}(6 - w) = 5 \cdot \frac{1}{5}(6 + w)$$

$$\rightarrow \quad 2(6 - w) = 1(6 + w)$$

$$\rightarrow \quad 12 - 2w = 6 + w$$

Subtracting 6 from both sides and adding $2w$ to both sides, we get $6 = 3w \rightarrow$ **w = 2 miles/hour**.

Step 5. We want *her speed on the return trip*, which according to the chart is $6 + w$. We know that $w = 2$, so her speed returning was $6 + 2 = $ **8**.

Answer: Kayla's return speed was 8 miles/hour.

What's the Deal?

How do we know which things to write in the chart and in what order? The truth is, there's often more than one order to do things in, and that takes a lot of pressure off! Specifically, you might fill up the chart *without* using all the important information the problem gives, and need another equation—which is fine! For example, in the Rollerblading problem above, if we hadn't used the fact that the total time is $\frac{3}{5}$ before we made the chart, here's how it might have gone. We knew the time going was twice the time coming back (which we called b), so we could have just filled in $2b$ and b for the times, and multiplied across:

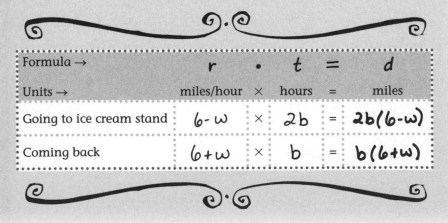

Formula →	r	•	t	=	d
Units →	miles/hour	×	hours	=	miles
Going to ice cream stand	$6 - w$	×	$2b$	=	$2b(6-w)$
Coming back	$6 + w$	×	b	=	$b(6+w)$

Then, because we knew the two distances are equal, we could have written the true statement **2b(6 − w) = b(6 + w)**. Now, taking the piece of information that is *not* reflected in our chart, namely, that the two times add up to $\frac{3}{5}$, we'd also have the second equation $2b + b = \frac{3}{5}$. Solving for b, we'd get $b = \frac{1}{5}$, and plugging it into the bolded equation above, we'd solve for w.

So now you see—there may be more than one way to do these problems, and just because you will use a piece of information doesn't mean it necessarily *has* to go in the chart. It's okay to leave it out and use it as a separate equation—and to experiment!

QUICK NOTE   Some people rearrange the formula $r \cdot t = d$ by solving for r or t, for example, $t = \frac{d}{r}$ or $r = \frac{d}{t}$. Sometimes they use that to help fill in the r or t columns. I just wanted to point it out in case you see it in class. But I'm not a fan, and honestly, it's not needed.

Hey, remember our guy in the field of waving grains? He's waiting patiently for us in problem 1 below, and we're now ready to tackle him. You know, figuratively speaking.

I recommend reading this whole chapter a second time, before trying these problems. This stuff can take some getting used to!

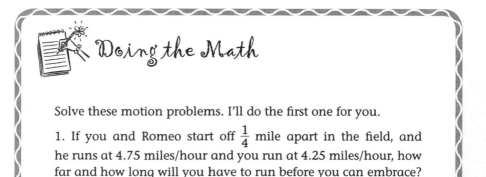

Doing the Math

Solve these motion problems. I'll do the first one for you.

1. If you and Romeo start off $\frac{1}{4}$ mile apart in the field, and he runs at 4.75 miles/hour and you run at 4.25 miles/hour, how far and how long will you have to run before you can embrace?

<u>Working out the solution</u>: The units look good and consistent; we've got miles and miles/hour. That means our answer will end up being in hours. Great! So, what do we know? Romeo's speed is 4.75 miles/hour, your speed is 4.25 miles/hour, and at the moment of your embrace, your *times* will be equal to each other. Let's start filling in a chart. Notice that by using only one variable for both columns, t, we are *saying* that the two times must be equal. (The two distances will not be equal.)

The <u>Moment</u> of Embrace:

Formula →	r	•	t	=	d
Units →	miles/hour	×	hours	=	miles
You	4.25	×	t	=	$4.25t$
Romeo	4.75	×	t	=	$4.75t$

Notice in our diagram that the distance you run plus the distance that Romeo runs will equal $\frac{1}{4}$ mile, so we can write this true statement,

You   Romeo

total distance $= \frac{1}{4}$ mile

$4.25t + 4.75t = \frac{1}{4}$, and then solve for t! Combining like terms, it becomes $9t = \frac{1}{4}$, so $t = \frac{1}{36}$ hours. To find the distance you've run, we plug $\frac{1}{36}$ for t into $4.25t$ to find out that you'd have to run

$$4.25t = 4.25\left(\frac{1}{36}\right) = \frac{4.25}{36} = \frac{425}{3600} = \frac{17}{144} \approx 0.12 \text{ miles,}$$

a little more than a tenth of a mile. For the answer, let's convert the time into minutes, and we get

$$\frac{1}{36} \text{ hour} = \frac{1}{36} \text{ hour} \cdot \frac{60 \text{ min}}{1 \text{ hour}} = \frac{60}{36} \text{ min} = \frac{5}{3} \text{ min} = 1\frac{2}{3} \text{ min}$$

or 1 minute and 40 seconds.

Answer: To reach Romeo, you'll run about 0.12 mile in 1 minute, 40 seconds. Not bad!

2. With help from the river's 2 mile/hour current, Jessica can ride her Jet Ski downstream in 12 minutes, and it takes her 18 minutes to ride her Jet Ski back upstream the same distance. How fast does the Jet Ski travel in calm water (with no current)?

3. Two trains leave the same station, traveling in opposite directions. One travels at 30 miles/hour and the other at 70 miles/hour. How long will it be until they are 50 miles apart? *(Hint: Make a chart for the* moment *they are 50 miles apart.)*

4. In still water, Amanda can row 3 km/hour. If it takes her 15 minutes to row upstream (against the current) from her campsite to the park, and only 3 minutes to row back down to the campsite, how fast is the current in the river?

5. Crystal leaves New York on a plane going 600 miles/hour to Las Vegas. At the same time, Dina leaves Las Vegas on a plane going 650 miles/hour to New York. If the cities are 2,500 miles apart, how far from New York are they when they can wave to each other as their planes pass? *(Hint: This is just like Romeo!)*

6. Erin and Devon ride their go-karts around a 150-meter circular track. If they start off together and Devon rides 50 meters/hour *faster* than Erin, how long will it be until Devon laps Erin? *(Hint: See the Kim/Claire example on pp. 188–90.)*

(Answers on p. 408)

Takeaway Tips

For motion problems, $r \cdot t = d$. Most of the time, you can make a chart for a round-trip or a particular moment in time.

Label the unknowns, use them to fill in the chart, and go from there.

Always pay attention to the units. They must be consistent.

TESTIMONIAL

Cecilia Aragon
(Berkeley, CA)
<u>Before</u>: Insecure outcast
<u>Today</u>: Daredevil airshow pilot and astrophysics computing scientist!

Growing up, I just never really believed I was "smart enough." I've struggled all my life with insecurities, and it was especially hard in middle school. In the small town where I grew up, we were one of the few Latino families. I remember walking into a store with my mother and seeing the clerk at the counter turn away and refuse to serve us. In school, even some of the teachers seemed biased against me. I was always put in the slow reading and math groups at the beginning of the year, and my low grades on essays and art projects often seemed unfair. My parents never really talked to me about racism, so I didn't understand what was going on.

"Growing up, I just never really believed I was 'smart enough.'"

But I realized that when you get an algebra answer right, *no one can take that away from you*, so I found myself working extra hard in math class. It was empowering to finally feel in control of my own success—no matter what anyone else thought of me.

I left for college (a much more diverse environment!), ended up majoring in math, and got a PhD in computer science. Now I'm proud to call myself an astrophysics computing scientist and daredevil airshow pilot. At Berkeley National Lab, I use math like geometry and calculus to study exploding stars and learn more about the fundamental nature of the universe. I also get to control powerful supercomputers. It's pretty awesome.

Believe it or not, airshow pilots need math in the middle of competitions! Using the airspeed and wind speed and the formula $r \cdot t = d$, I can estimate how far I will travel in a certain amount of time and make strategic, split-second decisions for when to begin complex maneuvers. Without that formula, I could easily get penalized for drifting outside the competition boundaries before the end of the trick.

One of my favorite maneuvers is called a torque roll. I point the airplane straight up, full throttle, until the engine can't overpower gravity anymore. Then slowly, I begin to slip backward through the air, rolling through the smoke from my plane. For awhile, the horizon whirls past me and I count the number of 360-degree rotations. Then—wham! The plane can no longer fly backward and flips over violently like a dart, and I'm flung to the limits of my seat belts, flying toward the ground faster and faster, reaching up to 200 miles per hour as the earth rushes toward me. I pull hard on the stick out of the vertical dive, up to 12 G's, or 12 times the force of gravity, as centrifugal force shoves me down into the seat, and I zoom just above the runway, ready for more. I love the thrill of it. And if it weren't for math, I could end up disqualified—or splattered on the runway!

I firmly believe that the reason more girls don't go into math is because they are not encouraged enough during those times when they get scared or insecure. But math can open up so many worlds to you, and trust me when I tell you that it's worth it to keep going and believe in yourself, even if no one else does.

That attitude made all the difference in my life— and it will in yours, too.

Chapter 14

Dip in the Pool, Anyone?

Rate of Work Problems

*Y*ou're throwing a pool party today, and all your friends are coming. It's gonna be a blast! What your parents didn't tell you was . . . the pool had to be drained this morning. Um hello, that would have been good information to have, but whatever.

The engineer just started refilling the pool using two pipes. But will your pool be filled in time? I mean, people start arriving in two and a half hours, and you haven't even made the hors d'oeuvres yet. Okay, deep breath. This is what the engineer tells you: "The smaller pipe, by itself, can fill the pool in 6 hours. The bigger pipe works twice as fast, and we've got both working together." Just then, she gets a phone call. Can you figure out how long until the pool is filled on your own? (Yes, you can!)

These problems involve a *rate of work,* whether it's filling pools, washing cars, or painting your nails. You might not believe me, but the key to these problems, and many of the types of word problems in this book, is that it's all about the *units.*

Ring Ring What's It Called?

Rate of work is the rate of some job being done, expressed as "jobs per time." For example, these are rates of work for washing puppies and painting houses:

$$\frac{3 \text{ puppies}}{30 \text{ min}}, \text{ or } \frac{1 \text{ puppy}}{10 \text{ min}}, \text{ or } \frac{1}{10} \text{ puppy per minute}$$

$$1 \text{ house in 4 hours, or } \frac{1 \text{ house}}{4 \text{ hours}}, \text{ or } \frac{1}{4} \text{ house per hour}$$

Jobs per Time

We should always write our rates as **jobs per time**, even if that means the "number of jobs" has to be a fraction, which it often will. For example, if it takes you 7 hours to paint one (huge!) banner, your *rate* is $\frac{1 \text{ banner}}{7 \text{ hours}}$, or $\frac{1}{7}$ **of a banner per hour**. Let's say you're really fast at washing cars, and you can wash 5 cars per hour. Your *rate* is $\frac{5 \text{ cars}}{1 \text{ hour}}$ or **5 cars per hour**. Then in 4 hours, you can wash 20 cars. That makes sense, right?

$$\underbrace{\text{5 cars/hour}} \cdot \underbrace{\text{4 hours}} = \underbrace{\text{20 cars total}}$$
$$\quad\quad rate \quad\quad\quad time \quad\quad total\ jobs\ done$$

Or in those same 4 hours, you could paint $\frac{4}{7}$ of a banner:

$$\frac{1}{7}\textbf{ banner/hour} \cdot \text{4 hours } = \frac{4}{7} \text{ of a banner total}$$

In fact, as long as the rate is expressed as "jobs per time," then:

rate · time = total jobs done

This equation will come up a lot—good to keep in mind!

So, back to our pool party crisis: The guests arrive in two and a half hours! The smaller pipe would take 6 hours to fill the pool by itself, and the bigger pipe works twice as fast, so it should take just 3 hours to fill the pool by itself, right? So, how long will it take them to fill the pool together? First, we need to find their *combined rate* of work.

Let's focus on the rates and write the smaller pipe's rate as *jobs per time,* or in this case, *jobs per hour*. What's the job? Filling the pool. So in this case, the rate is 1 pool per 6 hours, or $\frac{1}{6}$ **of a pool per hour**. That's the <u>rate</u> of the smaller pipe. With me so far?

The bigger pipe can fill 1 pool per 3 hours, so the bigger pipe's rate is $\frac{1}{3}$ **of a pool per hour**. Their combined <u>rate</u> is $\frac{1}{6}$ pool/hour + $\frac{1}{3}$ pool/hour = $\frac{1}{6} + \frac{1}{3} = \frac{1}{6} + \frac{2}{6} = \frac{3}{6} = \frac{1}{2}$ **pool/hour**. So, together, they can fill $\frac{1}{2}$ pool in 1 hour, which is the same as **1 pool in 2 hours**. Yes, the pool should be done way before people start showing up. Nice!

Answer: Working together, the pipes will fill the pool in 2 hours.

QUICK NOTE   If we'd been asked how long it would take these pipes to fill a total of 5 pools, we'd continue from here by writing the equation from p. 199, filling in that the rate is $\frac{1}{2}$:

$$\text{rate} \cdot \text{time} = \text{total jobs done}$$
$$\frac{1}{2} \cdot \text{time} = 5$$

And solving, we'd multiply both sides by 2 and get that the time = **10 hours.**

Watch Out!

Notice that the key to solving the pool problem was writing the pipes' rates as <u>jobs per time</u>: $\frac{1}{6}$ and $\frac{1}{3}$. You might be tempted to do something like add 6 and 3, but these are not rates—they are *times*! Adding 6 + 3 = 9 would represent, I don't know, the number of hours it would take if one pipe filled the pool, the pool was instantly drained, and then the other pipe filled the pool. Yikes! <u>If you remember only one thing from this chapter, I want you to remember to write the rates as jobs/time.</u> Everything works out if you start that way.

QUICK NOTE   The "jobs per hour" rates can take a few forms, for example:

$$\frac{5}{4} \text{ jobs per hour} = \frac{5}{4} \text{ jobs/hour} = \frac{\frac{5}{4} \text{ jobs}}{1 \text{ hour}} = \frac{5 \text{ jobs}}{4 \text{ hours}}$$

These are equivalent expressions; they mean the same thing.

Let's put this down in a step-by-step method. Often, these problems will require using variables, so I'll include that in the steps.

Step By Step

For "rate of work" word problems:

Step 1. Look at the units you are given. Then, <u>write all the *known* rates in terms of "Jobs per time"</u>—usually "jobs/hour" or "jobs/min." Then label the *unknown* rates. You can just give them letters, or if it's easy, write them in terms of the *known* rates. Make sure you're dealing in consistent units (hours or minutes, miles or feet, and so on).

Step 2. Write down any relationships you know, using your variables. For example, add up the combined rate, if that's what the problem involves.

Step 3. You might be done already. But if not, then use the information given to write a true statement in terms of just *one* variable, perhaps using **rate · time = total jobs done**.

Step 4. Solve for the variable. This could be rate, time, or total jobs done.

Step 5. Make sure you've solved for what the problem was asking for. If not, use what you've done to solve for that value. And remember, keep your units consistent.

QUICK NOTE When rates are written as "jobs per time," you can add and subtract them as needed.

And... Action! Step By Step In Action

You just realized you're supposed to wash your parents' car before the party! Luckily, your cousin's going to help. You've washed it together before, and it takes 10 minutes. You've also washed it by yourself, which takes 15 minutes. Since you have to make the hors d'oeuvres, your cousin offered to wash the car by herself. How long will it take her?

Step 1. We have consistent units: <u>minutes</u> and <u>cars</u> washed. Writing our rates as **jobs/time**, your car-washing rate is $\dfrac{1 \text{ car}}{15 \text{ min}}$, or equivalently, $\dfrac{1}{15}$ **of a car per minute**. Also, your *combined* rate is $\dfrac{1}{10}$ **car per minute**. Let's call your cousin's rate c.

Steps 2 and 3. The rates are in jobs/time format, so we can write this true statement about the combined rate: $\dfrac{1}{15} + c = \dfrac{1}{10}$. There's our equation!

Step 4. Solve for your cousin's rate: $\dfrac{1}{15} + c = \dfrac{1}{10} \rightarrow c = \dfrac{1}{10} - \dfrac{1}{15} = \dfrac{3}{30} - \dfrac{2}{30} \rightarrow c = \dfrac{1}{30}$. Her rate is $\dfrac{1}{30}$ of a car in 1 minute; in other words, she washes 1 car in **30 minutes**. She's just really careful, that's all (or maybe it's all that texting . . .) And hey, we're done!

Answer: 30 minutes

QUICK NOTE   If you ever feel lost and aren't sure how to find what's being asked, start by solving for an unknown *rate*. That is often the key. These are *rate* of work problems, after all!

Take Two: Another Example

You're back at the pool, you've made your hors d'oeuvres, you've got your cute tankini and cover-up on, and it's an hour until the party starts. You go outside to check the pool's progress. But the pipes just stopped working after filling just $\dfrac{3}{4}$ of the pool. Argh!! They brought in the neighbor's hoses, which just started filling the pool again.

Now the engineer (who doesn't seem to "get" the gravity of the situation) says, "Well, <u>if we had all 3 hoses going at once, they could fill an entire pool in 2 hours</u>. But the largest hose is broken. The medium-sized hose fills a pool twice as fast as the smallest, and the large-sized *hose fills a pool three times as fast as the smallest." Oops, there goes her cell phone again. Who is she talking to, anyway?*

Again, it's up to us to figure it out: If it would take all three hoses 2 hours to fill an entire pool but the largest hose is broken, how long will it take the small- and medium-sized hoses, working together, to fill just the last $\frac{1}{4}$ of the pool?

Step 1. Okay, a monster problem, but we'll take this a step at a time. Our units are all hours and pools being filled. What rate(s) do we know? We *know* that the combined rate of all three hoses would be 1 pool in 2 hours—in other words, $\dfrac{1 \text{ pool}}{2 \text{ hours}}$, or $\frac{1}{2}$ pool per hour. Let's label the unknown rates for the small, medium, and large hoses as s, m, and b. (I'm using b for big, because I don't like l as a variable. It looks too much like the number 1.) And notice that if we find the combined rate $s + m$, then we'll know <u>the rate of the two working hoses</u>. This rate will be the key to figuring out how long it will take them to fill the rest of the pool. (Nobody said this stuff was easy, but stick with me!)

Step 2. Well, we've been told that the combined rate of *all* the hoses is $\boldsymbol{s + m + b = \frac{1}{2}}$, right? What other information do we know? Hmm— we've been told that the medium hose's rate is twice the smallest, so that translates into math like this: $\boldsymbol{m = 2s}$.* We also know that the largest hose works three times as fast as the smallest, so that means $\boldsymbol{b = 3s}$.

Step 3. Now, all the hoses' rates are expressed in terms of one variable, which is great! So we can rewrite our big equation in terms of s.

$$s + m + b = \frac{1}{2} \text{ has now become } \boldsymbol{s + 2s + 3s = \frac{1}{2}}$$

Step 4. Solve for s! So combining like terms, we get $6s = \frac{1}{2}$. Dividing both sides by 6, we get $\boldsymbol{s = \frac{1}{12}}$. What is $\frac{1}{12}$? It's the rate of the smallest hose, which means it can fill $\frac{1}{12}$ of a pool in 1 hour. We know that $m = 2s$, so the medium hose's rate is $m = 2s = 2\left(\dfrac{1}{12}\right) = \dfrac{1}{6}$.

. . . .

* I like to plug in numbers, like 2 & 1, just to make sure I got the order right and that it shouldn't have been $2m = s$. It's just a good reality check. Careless mistakes are so easy to make, but some (like this kind) are easy to prevent.

Step 5. To solve for the *time* it takes for the small and medium hoses to fill the last $\frac{1}{4}$ of the pool, we first need their combined *rate*: $\frac{1}{12}$ pool/hour $+ \frac{1}{6}$ pool/hour $= \frac{1}{12} + \frac{1}{6} = \frac{3}{12} = \frac{1}{4}$ **pool/hour**.

We only have $\frac{1}{4}$ of the pool that still needs water, so we know we'll be done in **1 hour**. Yay, we're just barely going to make it! As long as the engineer gets out in time . . . but she seems pretty social, maybe we should invite her?

Answer: The small and medium hoses will fill the last quarter of the pool in 1 hour.

We're going to make sure you're never intimidated by any of these 'ol work problems. Get strong muscles here, and you can handle anything those teachers try to throw at you. Remember: Jobs per time!

"When I do something difficult in math, my brain kind of hurts sometimes. But when I stick with it, then I get smarter and learn something new." **Briana, 14**

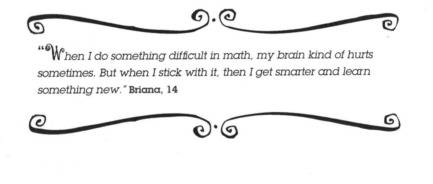

Doing the Math

Solve these rate of work problems. I'll do the first one for you.

1. Amanda and Michael are making dessert for the local homeless shelter. Amanda can bake a cake in 2 hours. Michael can bake 2 cakes in 5 hours. Baking together, how long will it take to bake 12 cakes?

<u>Working out the solution</u>: In "jobs/hour," Amanda's rate is $\frac{1}{2}$ of a cake per hour, and Michael's rate is $\frac{2}{5}$ of a cake per hour. So combining their rates, we get $\frac{1}{2} + \frac{2}{5} = \frac{5}{10} + \frac{4}{10} = \frac{9}{10}$. And what does this mean? It means that together, they can bake $\frac{9}{10}$ of a cake in 1 hour. Now that we have the rate, in order to find the *time* it takes to bake 12 cakes, let's use our **rate · time = total jobs** equation. Using "12 cakes" as the total jobs, we'll solve for time: $\frac{9}{10} \cdot t = 12$.

Multiplying both sides by 10 and dividing by 9, we get $t = \frac{120}{9} = \frac{40}{3} = 13\frac{1}{3}$ **hours.** What's $\frac{1}{3}$ of an hour? Why, 20 minutes, of course!

Answer: It will take them 13 hours and 20 minutes to bake 12 cakes.

2. Megan and Katie decided to sell handmade bracelets. Katie can make 3 bracelets in four hours, and Megan can make 4 bracelets in five hours. Working together, how long until they have made 31 bracelets?

3. Fia has bought her little sister a dollhouse, and she and her dad are going to hand paint it. By herself, Fia could paint it in six hours. By himself, her dad could paint it in four hours. Working together, how long until they complete this one job?

4. Lily's mom can build a snowman in 2 hours. Together, Lily and her mom can do it in 45 minutes. How long does it take Lily to build a snowman by herself? *(Suggestion: Convert the hours to minutes, get the answer, and then try solving it using hours.)*

5. When four sisters work together, they can decorate 20 invitations per hour. Jen is twice as fast as Sarah, Emily is three times as fast as Sarah, and Hillary is 4 times as fast as Sarah. **a.** What is Sarah's rate? **b.** Without Hillary, how long would it take them to make 18 invitations?

6. It takes 8 minutes to fill Carly's bathtub, and it takes 12 minutes to drain it. Carly can't find the plug anywhere, but she just came back from skiing and really wants to warm up in a nice hot bath. With the water on but the drain open, how long does it take for the tub to be filled?* *(Hint: Express the drain rate as a negative rate: $-\frac{1}{12}$ jobs/min.)*

(Answers on p. 408)

What's the Deal?
Unit Multipliers and Rates

This section will open your eyes and help you with all sorts of word problems.

From unit conversions,[†] we learned about **unit multipliers**—<u>fractions that equal 1</u> (much like copycats) but that use *different units* on top and bottom, like $\frac{12 \text{ inches}}{1 \text{ foot}}$ or $\frac{60 \text{ sec}}{1 \text{ min}}$. For example, we can convert 180 seconds by multiplying it times $\mathbf{\frac{1 \text{ min}}{60 \text{ sec}}}$ to get:

$$180 \text{ sec} = \frac{180 \text{ sec}}{1} \cdot \frac{\mathbf{1 \text{ min}}}{\mathbf{60 \text{ sec}}} = \frac{180}{60} \text{ minutes} = \mathbf{3 \text{ minutes}}$$

The reason we chose $\frac{1 \text{ min}}{60 \text{ sec}}$, and not $\frac{60 \text{ sec}}{1 \text{ min}}$, is because we want the "seconds" to cancel from the top and bottom of the fraction, and what we're left with is the number of minutes *equivalent* to 180 seconds. Similarly, if our car-washing rate is 1 car in 20 minutes, then $\frac{1 \text{ car}}{20 \text{ min}}$ can be thought of as our "unit multiplier." Washing cars for 3 hours (180 min):

$$\frac{1 \text{ car}}{20 \text{ min}} \cdot 180 \text{ min} = \frac{1 \text{ car}}{20 \text{ min}} \cdot \frac{180 \text{ min}}{1} = \frac{180}{20} \text{ cars} = \mathbf{9 \text{ cars}}$$

. . . .

* Don't ever waste water like this!
† For a full review of unit conversions, see Chapter 19 in *Math Doesn't Suck*.

The minutes cancel away, and we're left with *cars*—the number of cars "equivalent" to 3 hours of work. Kind of interesting, huh? In general, this is what's going on:

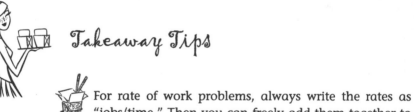

$$\frac{\text{jobs}}{\text{time}} \cdot \text{time} = \frac{\text{jobs}}{\cancel{\text{time}}} \cdot \frac{\cancel{\text{time}}}{1} = \text{total jobs done}$$

And similarly, the reason the equation $r \cdot t = d$ works (we used this in Chapter 13) is because the *time* units cancel. After all, speed is just distance/time:

$$\frac{\text{distance}}{\cancel{\text{time}}} \cdot \frac{\cancel{\text{time}}}{1} = \text{total distance traveled}$$

Kinda cool, right? When the units cancel, we know we're on the right track!*

Takeaway Tips

For rate of work problems, always write the rates as "jobs/time." Then you can freely add them together to get combined rates.

Make sure your units are consistent, and do conversions when necessary.

Remember: 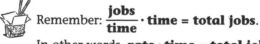 $\dfrac{\text{jobs}}{\text{time}} \cdot \text{time} = \text{total jobs}$.
In other words, **rate · time = total jobs**.

If you can't figure out how to start, see what *rates* you know, and use them to find the unknown rate(s). That's usually the key to finding whatever the problem wants.

. . . .

* For more on the versatility of unit cancellation, check out danicamckellar.com/hotx.

Chapter 15

Savvy Business Chick

Word Problems: Percents and Simple Interest

$\mathcal{H}$ave you ever dreamed of being a savvy Wall Street gal? Ever dreamed of owning your own business someday? Well then, you're in luck—because this chapter covers percents, profits, and simple interest—some great tools to get you on your way!

Percents are really important, especially for understanding money. You know, just that small detail in life that makes the world go 'round. So why not be an expert?

First, let's do a quick review of percents.*

Quick Percents Review

To convert a percent to a decimal: Take off the % sign and move the decimal point two places to the left. Examples: 25% = 0.25; 70% = 0.7; 7% = 0.07; 0.02% = 0.0002

To convert a decimal to a percent: Move the decimal point two places to the right and stick on a % sign. Examples: 0.5 = 50%; 0.05 = 5%; 1.2 = 120%; 0.001 = 0.1%

To convert a fraction to a percent: Convert the fraction to a decimal and then to a percent.

. . . .

* For an in-depth review of percents, check out Chapters 13–15 in *Math Doesn't Suck*.

Percent Increase/Decrease

Say a guy walks up to you—let's call him Bob—and he wants you to invest your money with him. He says, "Invest $100 with me. The first year, your money will decrease by 60%, but the next year it will increase by 80%. Trust me, you'll end up ahead!" Is this as good as it sounds? Should you trust this "Bob" fellow?

Besides the fact that he's wearing a used-car-salesman jacket, no, you should not trust him! Bob's either trying to swindle you, or he doesn't understand percents and should definitely not be handling anyone's money. Let's see how these percent problems work, and then we'll revisit this guy and his offer.

Like most word problems, percent problems come down to collecting the important information and translating it into math. Generally speaking, to see how much money we'll have after we've increased it or decreased it by a certain percent, we'll do:

$$\textbf{New Amount} = \boldsymbol{P} + \boldsymbol{Pr}$$

or

$$\textbf{New Amount} = \boldsymbol{P} - \boldsymbol{Pr}$$

Where *P* is the principal amount you started with and *r* is the % increase or decrease, written as a decimal (often called the interest *rate*, which is why we use *r*).

Let's say a $30 pair of shoes is discounted 10%. Because 10% of $30 is $3, we'll subtract off $3, right? Sure enough, look at what the formula tells us. (Remember: 10% = 0.1):

$$\text{New Amount} = P - Pr$$

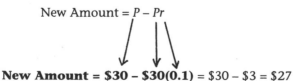

$$\textbf{New Amount} = \$30 - \$30(0.1) = \$30 - \$3 = \$27$$

QUICK NOTE I don't want you to think of this as a formula to memorize. In fact, I only want you to think to yourself, *Ah yes, I'll end up with the money I started with, plus or minus a percent of that money, and that is the formula.* It's important to keep your brain on, because blindly memorizing simple formulas without understanding them is bad—very bad. You'll thank me later.

Step By Step

Solving percent increase/decrease word problems:

Step 1. Translate English into math, using a variable for the unknown. Remember, if "of" is immediately surrounded by two values, that will translate into multiplication. Usually we'll end up with something like:

$$\text{New Amount} = P + Pr \quad \text{or} \quad \text{New Amount} = P - Pr$$

where P is the previous amount and r is the % increase/decrease in decimal form.

Step 2. Make sure you've written percents as decimals, so you can multiply them!

Step 3. Solve for the unknown, and repeat Steps 1–3 as necessary.

Step 4. Do a reality check. Does your answer seem reasonable? If so, you're done!

And... Action! Step By Step In Action

Cristi's got $30 to spend on a birthday present. She walks into a store that is having a sale: 20% off the current marked price! So, what's the highest marked price she can afford?

(Read this solution twice, all the way through. It'll make more sense the second time.)

Steps 1 and 2. Let's see: The highest *marked* price Cristi can afford would be the price which, when 20% is taken off, costs exactly $30, right? So the "New Amount" is $30, and let's label the *marked* price P—that's what we'll solve for. The math sentence we want to say is "$30 <u>is equal to</u> the previous price after a 20% discount." Agreed? Here's how that looks in math language:

$$\$30 = P - 0.2P$$

what she pays ↗   Marked price ↑   ↖ discount off <u>Marked</u> price

Step 3. Now we just solve for P. Combining like terms, isolating P, and using the copycat* $\frac{10}{10}$ to get rid of decimals, we get:

$$\$30 = 1P - 0.2P$$

$$\rightarrow \$30 = 0.8P$$

$$\rightarrow P = \frac{30}{0.8}$$

$$\rightarrow P = \frac{30 \times \mathbf{10}}{0.8 \times \mathbf{10}} = \frac{300}{8} = \frac{75}{2} = 37.5.^{\dagger}$$

In dollars, that's **$37.50**.

Step 4. Reality check: Does a marked price of $37.50 make sense? It does seem like after 20% is taken off, it would only cost about $30. (If we'd gotten an answer like $375 or $3.75, we'd know something had gone wrong!) We could also check our work by finding 20% of $37.50, which is $7.50, and subtracting it off $37.50. Yep, we get $30! So Cristi can afford to buy anything that is currently marked as $37.50. Nice.

Answer: $37.50

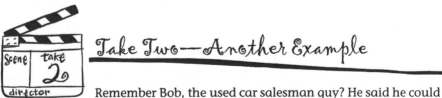

Take Two—Another Example

Remember Bob, the used car salesman guy? He said he could invest your $100 and that it would first decrease in value by 60% but it would then increase in value by 80%. Let's see what would happen.

· · · · · · · · · ·

* For more on copycat fractions, see Chapter 4.
† We also could have divided 30 ÷ 0.8 = 37.5.

We'll do this in two stages. First, we'll only look at what happens when the $100 decreases by 60%. You might say, "That's easy. Just subtract $60 from $100 and you'd get $40." And you would be right, but then it gets a little trickier. Let's do this step by step, and you'll see what I mean.

Steps 1 and 2. Starting with $100, if we want $100 decreased by 60% (which equals 0.6 in decimals), then our new amount = $100 – $100(0.6).

Step 3. Simplifying, we get $100 – $100(0.6) = $100 – $60 = **$40**. With me so far? At this point, we start over again with Steps 1 and 2, with your $40. You can forget about your $100. It doesn't exist anymore. Notice that increasing your $40 by 80% does *not* mean increasing by $80. It means increasing by 80% of $40, and 80% equals 0.8, so:

$$\text{New Amount} = \$40 + \$40(0.8) = \$40 + \$32 = \mathbf{\$72^*}$$

Step 4. It's kind of hard to do a reality check on this kind of multistep problem because the results are so counterintuitive, but you can do it in stages. Does it make sense that 60% off $100 would be $40? Yes. And does it also makes sense that increasing $40 by 80% would give $72? Sure. (Note that increasing $40 by 100% would give $80.)

Answer: If you give him $100, you'll end up with only **$72**.[†] It's a bad deal. Run!

Watch Out!

Make sure to do these multiple increase/decrease problems in stages. It would be easy to think the $100 would end up being $120 if you read the problem quickly, thinking, "Oh, that's just $100 – $60 + $80 = $120." But now you know why that's wrong! The second increase or decrease <u>is based on the intermediate number</u> that you get—not on the first number. After we get the intermediate amount of money, we must pretend we're starting over fresh, like we never even saw the first amount of money. It's history!

· · · · · · · · · ·

* To review decimal multiplication, check out Chapter 10 in *Math Doesn't Suck*.
[†] Believe it or not, we'd get the same answer if we first increased $100 by 80% and then decreased it by 60%. Try it!

Reality Math

"Take an additional 20% off this 50% off sale!"

Is this as good as a 70% off sale? Sure seems like it, but don't be fooled . . .

Imagine that a dress originally cost $100; you might think you're about to get it for $30. But check it out: If it was first marked down by 50%, it would cost $50, right? Now if we take an additional 20% off of $50, we'd be paying $50 – $50(0.20) = $50 – $10 = **$40**. That's **60% off**, not 70% off. (Notice that the 20% was multiplied by $50, not $100.)

Savvy shoppers, unite!

Doing the Math

Solve these percent word problems. I'll do the first one for you.

1. Clare's going into the fashion business. She bought a bunch of dresses and is now selling them for $33 each, which is giving her a 20% profit.

    a. How much did she originally pay for each dress?

    b. If she wants to make a 30% profit, how much should she be charging instead?

<u>Working out the solution</u>: Let's label P as the price she originally paid for each dress. And $33 <u>is</u> 20% more than she paid for each dress, right? In math language, that's $33 = P + P(0.2)$. Combining like terms, we get $33 = P(1.2)$. Dividing both sides

by 1.2 and then simplifying, we get $P = \dfrac{33}{1.2} = \dfrac{33 \times 10}{1.2 \times 10} = \dfrac{330}{12} = \dfrac{55}{2} =$ **$27.50**. That's how much she originally paid.

Part **b** asks us how much she should charge to make a profit of 30%. So since she originally paid $27.50 per dress, we should find out what a 30% increase on <u>that</u> price would be. No problem! New Amount = $27.50 + $27.50(0.30) = 27.5 + 8.25 = **$35.75**. Does it seem reasonable that if she paid $27.50 per dress, then charging $33 would mean she made 20% profit, and charging $35.75 would mean she made 30% profit? It sure does!

Answer: a. She paid $27.50. b. She would have to charge $35.75 to make 30% profit.

2. Matthew invested $50 in the stock market. The first year, it shrunk by 10%. The next year, it grew by 10%. How much does he have now?

3. Caroline invested $50 in the stock market. The first year, it grew by 10%. The next year, it grew by 10% again. How much does she have now?

4. Sabrina is selling small handbags for $6 each.

    a. If she's making a 50% profit, how much did she pay for each handbag?

    b. If she wanted to make a 75% profit, how much should she charge instead?

5. Meghan invested in stocks. Unbelievably, her portfolio grew by 25% this year! She now has $100. How much did she start with?

6. Cherise invested $500 in the stock market. In the first year, it shrunk by 20%. In its second year, it increased by 30%, and in its third year, it decreased by 10%. How much does Cherise have now? *(Hint: Do this in three stages.)*

(Answers on p. 408)

What's the Deal?

If you did problem 2 above, you might have been surprised at the answer. It seems crazy. How can you *decrease* an amount of money and then *increase* it by the <u>same percent</u>, and end up with less than you started with! I mean, right? The key is to understand that <u>percents are only meaningful when they are percents OF some amount</u>. And when that amount changes, so does the value of the meaning of that percent.

You see, if we first decrease an amount by 10%, now it's a smaller amount of money. So, when we increase it by 10% (<u>of</u> that new, smaller amount), our *increase* won't be as big a number as our decrease was . . . and we end up with less than we started with.

Similarly, if we instead increase an amount by 10%, now we have more money. But then when we decrease that money by 10% (<u>of</u> that higher amount), that *decrease* will be a bigger number than the increase was, and we end up with less than we started with.

Read that again a few times until it sinks in.

When you hear about the stock market fluctuating up and down, now you understand how people can lose so much—and some of them don't realize that hidden in that fluctuation is a loss of money. The more volatile (up and down) the stocks are, the more people lose. For more on how this works, check out danicamckellar.com/hotx.

Being a savvy investor with your hard-earned money is a real benefit of understanding algebra, and it's making you a more powerful young woman, right now.

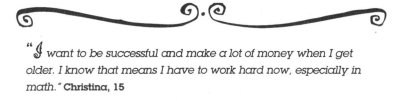

"*I want to be successful and make a lot of money when I get older. I know that means I have to work hard now, especially in math.*" **Christina, 15**

Simple Interest

One type of "percent increase" problem involves **simple interest**. Banks pay their customers a small amount of interest to keep money at their bank, say at a yearly rate of 2%.* That means at the end of one year, if you don't take any money out, the value of your bank account will increase by 2%. For example, if you put $100 in a bank account bearing 2% interest, after one year you'd have:

$$\text{New Amount} = \$100 + \$100(0.02)$$

$$= \$100 + \$2 = \mathbf{\$102}$$

Just like we've been doing, right? Now, suppose you took the money out early, say after just 3 months ($\frac{3}{12}$ of a year). Then all you'd have to do to calculate how much you have is multiply the <u>yearly</u> interest rate (0.02) by the <u>fraction of the year</u> you kept the money in the account, $\frac{3}{12}$. So if you kept your $100 in this account for only 3 months, you'd end up with:

$$\text{New Amount} = \$100 + \$100(0.02)\left(\frac{3}{12}\right)$$

$$= 100 + 2\left(\frac{1}{4}\right) = 100 + 0.5 = \mathbf{\$100.50}$$

Hmm, 50 cents, eh? Not much profit after 3 months. Then again, neither was $2. I guess the bank doesn't feel like giving very much money away. Who knew?

Ring Ring ✦ **What's It Called?**

Simple Interest is the interest income (money) made after a particular amount of *time*, calculated by the following formula:

$$i = Prt$$

. . . where i = interest <u>i</u>ncome, P = <u>P</u>rincipal (money we started with), r = interest <u>r</u>ate, and t = <u>t</u>ime. The longer you invest the money at this rate, the more money you will make.

. . . .

* If you're curious, the reason banks pay people to keep their money is because banks actually *use* that money to give *other* people loans for houses and such, and banks charge those other people a much higher interest rate.

For example, let's say you worked for a summer, saved up $1000, and opened a bank account—good job! How much interest would you make on a bank account yielding (producing) 4% yearly interest on a principal amount of $1000 for 2 years?

Since $P = 1000$, $r = 0.04$, and $t = 2$, we get:

$$i = (1000)(0.04)(2) = 40(2) = 80$$

So we'd make $80 in interest after two years. Notice that if you kept the money in the account for just one year, you'd make $40. However, if you kept the money in for 3 years, you'd make $120. And that makes total sense; you're making $40 *per year*.* Assuming of course, that you didn't take any money out of the account that whole time . . .

 QUICK NOTE   Notice that the formula $i = Prt$ only gives us the amount of <u>interest</u> money we made, not the new total money like we found with percent increase/decrease problems. If you want the new *total* amount, add the Principal Amount to the interest income, just like we did before, only this time we're taking into account the *time* with the t variable:

$$\text{New Amount} = P + \underbrace{Prt}_{i}$$

Make sense? Remember what I told you: Don't think of these as formulas. I want you to think about what they mean, and then you'll never be confused!

. . . .

* FYI, in the real world, you would make a bit more money than this because banks use *compound interest* instead of simple interest. (More on that in Algebra II.)

Watch Out!

Make sure the *rate* and *time* have consistent units. This will usually be years, because rates are usually *yearly* interest rates, that is, a percent increase *per year*. So *t* should be a number of years. Now, if you want to know how much simple interest you'd make in just one month, you'd use $\frac{1}{12}$; for 3 months, you'd use $\frac{3}{12}$, or equivalently, $\frac{1}{4}$; and for one day, you'd use $\frac{1}{365}$.

Step By Step

Simple interest problems:

Step 1. Figure out what you know and what the problem is asking for, and check your units. Unless stated otherwise, the interest rate means interest per *year*. Convert the time, *t*, to years if necessary.

Step 2. Use the simple interest formula $i = Prt$, where i = interest income (the money you make), P = the principal amount (money you started with), r = the interest rate (as a decimal), and t = the time you're investing the money for in years.*

Step 3. Reality check—does your answer make sense? If so, you're done!

Watch Out!

Make sure to label your variables correctly. After all, the word "interest" comes up a lot! In simple interest problems, when you see a percent, that's the interest <u>rate</u>: *r*. The variable *i* usually stands for the interest <u>income</u>: an amount of money.

. . . .

* Heads up: The only time you'd use a unit other than years is if the interest rate isn't given in terms of years, but it almost always is.

Susan bragged that she put $40 in a bank account, and after one month it was worth $70. What would that bank's yearly interest rate have to be? Is she, like, lying or what?

Step 1. Looks like we'll be solving for the interest <u>rate</u>: r. So, how much interest did Susan supposedly make? If she ended up with $70, that means she made $30 in interest, so $i = \$30$. The principal amount of money she started out with was $40, so $P = \$40$, and what is t? One month. Because the problem asks for a yearly interest rate, the t must be expressed in terms of years, so $t = \frac{1}{12}$.

Step 2. Using our formula, $i = Prt$, we get:

$$30 = (40)\left(\frac{1}{12}\right)r$$

$$\rightarrow \ 360 = 40r$$

$$\rightarrow r = \frac{360}{40} = \frac{36}{4} = \mathbf{9}$$

Step 3. Ah, you might think, 9% sounds like a really good interest rate for a bank. But hold the phone; remember, 9 is currently expressed as a *decimal*. What does that translate into as a percent? Move that decimal to the right two places, stick on a %, and you get 900%. Um, wow. I've never heard of a bank giving 900%. Susan's totally lying.

Answer: The interest rate would have to be 900%.

Doing the Math

Assume that all banks in these problems use simple interest. I'll do the first one for you.

1. If John put $50 in the bank at a 5% yearly interest rate, how long must he wait before his account is worth $60?

Working out the solution: The interest income he makes will be $10, right? We know that $i = 10$, $P = 50$, $r = 0.05$, and we'll solve for t. So, $i = Prt$ becomes $10 = 50(0.05)t$. To multiply 50 (0.05), just multiply 50 times 5, and then put two decimal places back in for the answer: $250 \rightarrow 2.5$, so, $10 = 2.5t$. Divide both sides by 2.5, and then

$$t = \frac{10}{2.5} = \frac{100}{25} = 4. \text{ Done!}$$

Answer: John must wait 4 years.

2. How much interest income is made with an investment of $400 for 3 months at a 2.5% yearly interest rate?

3. If the bank is offering a 3% yearly interest rate, how much must you put in the bank to end up with $195 in interest income after $6\frac{1}{2}$ years?

4. If Laura put $50 in the bank at a 5% yearly interest rate, how long must she wait before her entire account is worth $110?

(Answers on p. 409)

What's the Deal?

How can we have *t* in this "simple interest" formula $i = Prt$, but we never saw *t* in the percent increase/decrease formula from p. 209 in this chapter? That's a very good question. Earlier in this chapter, we treated every increase or decrease like it happened just once. It's the same as if the percent increase/decrease happened *per year*, and we calculated what happened after *one year*. In that case, the time units match, and $t = 1$. Or, we could have been pretending that the rate of increase/decrease happened *per day*, and we calculated what would happen after *one day*. Again, because the units match, $t = 1$. So you see, *t* has been with us all along; we just couldn't see it because it was an invisible 1!

Takeaway Tips

 Percents are only meaningful when they are percents "of" some value.

 When an amount, *P*, has a % increase or decrease, the resulting amount will be $P + Pr$ or $P - Pr$, where the percent, *r*, is expressed as a decimal.

For multiple increase/decrease problems, solve them in stages. Remember: The second increase or decrease <u>is based on the intermediate number</u> that you get.

Simple interest is calculated with this formula: $i = Prt$, where $i =$ interest income, $P =$ the starting amount of money, $r =$ the interest rate, and $t =$ the time the money is invested. Make sure the units always match.

TESTIMONIAL

Tracy Trump
(North Manchester, IN)
<u>Before</u>: Suffered from medicine-induced depression
<u>Today</u>: Executive director for charities!

I've always been competitive—in my grades, sports, you name it. My prayers often included a plea to allow us to win sectional in basketball, and I admit that I enjoyed getting a higher math grade than the rude boy who sat beside me. But during my junior year of high school, I suddenly developed a lung disease that brought my life to a standstill: I ended up in the hospital multiple times (including many emergency room visits, emergency chest tubes, and two surgeries) and missed several weeks of school. I was in pre-calculus at the time, which was already challenging, but learning it on my own made it nearly impossible. To make things worse, I had to take medical steroids for more than a year, which altered my moods for the rest of high school. I struggled with depression, anger, and anxiety, and interacting with even my best friends became a major challenge. Then I also developed "moon face"—a bloated, swollen appearance—from the steroids. I lost all confidence in myself and my abilities, which made schoolwork even more difficult, especially pre-calculus. But I kept studying, no matter how depressed I felt. It was the hardest, loneliest time in my life.

> "I lost all confidence in myself and my abilities."

It was during this struggle and in my recovery that I realized that God had more important things in mind for me than winning sectional or getting a better grade than the boy sitting next to me. Eventually, I was able to redirect my anger and anxiety toward becoming a better student and person as a whole, but it took time. It helped that when I returned to school, I had the best math teacher in the world, Lucy Dundore. She obviously loved math and loved to teach others about it, too. She was never too busy or tired of explaining theories or algorithms to me.

Today, I am totally healed and still super competitive, but I better understand the importance of faith, family, friends, love, and helping each other in difficult situations. And now I help people for a living; I manage money for charities!

I'm the executive director of a nonprofit organization that manages 230 endowed funds for other charities, and I use math every day. A way to think of an endowed fund is that someone gives us money (a principal amount) and says, "I want you to put this money in a bank account that makes interest. Don't ever spend the principal, but take all the _interest_ every month, and give it to this children's charity, forever." It's my job to make sure this money gets invested properly (in a bank or some other income-generating entity) and then to make sure its interest income is distributed to the charities correctly, so I often use the formula you ladies learned on p. 217:

$$i = Prt$$

In an endowment, the interest income, i, goes to the charities, while the P stays invested. That way the charity can keep getting money and _not ever run out_. I love knowing that I can make these generous donations grow into even larger amounts that can help even more. No matter what I was going through in high school, I always worked my butt off in math class. Now I'm grateful to say that when I work hard in math, I'm helping charities—and it feels great!*

. . . .

* Danica says, _Wanna help a charity while doing math? Check out mathathon.org!_

Chapter 16

The Perfect Party Snack

Word Problems Involving Mixtures

J love organic raw almonds and dried cranberries; they make a great healthy snack! In fact, it's one of the snacks I request when I'm filming movies or TV shows, but I also eat it at home. It has protein, it's a good energy food, and it's super tasty.

Let's say you're having a party and you want to get a total of 3 pounds of this mixture for your guests. If raw almonds cost $5.50 per pound and cranberries cost $7.50 per pound, how many pounds of each should you buy if you want to spend $20 total?

Ah, mixture problems. They can seem so confusing at first, but soon they'll be a breeze. In this chapter, we'll look at all sorts of mixture problems dealing with money, liquids, and more! Speaking of money . . .

Change Purse

If you have a wallet with a change purse, then you probably have a few quarters and dimes in it. To quickly find out how much money you have from the quarters and dimes, we could use this formula:

$$0.25q + 0.10d = \text{total money (in dollars)}$$

money from quarters   money from dimes

. . . where q = the *number* of quarters and d = the *number* of dimes. So if we had 3 quarters and 2 dimes, we could use the above formula and get

$0.25(3) + 0.10(2) =$ **0.95 dollars**. This makes sense because 3 quarters and 2 dimes is just: 75¢ + 20¢ = 95¢.*

Now, what if somebody has three times as many dimes as quarters and the entire thing adds up to $3.30? Could we figure out how many of each coin she has?

Yep! We'll use that same formula, $0.25q + 0.10d =$ total money. Because we know the total money is $3.30, we can write **0.25q + 0.10d = 3.30**. And because there are 3 times as many dimes as quarters, we also know that $d = 3q$, right? (Remember, d and q are the *numbers* of each coin we're holding.)

Now we have two equations in two variables, which we can solve![†] Let's use the smaller equation, $d = 3q$, to substitute $3q$ for d into the bigger equation, and we'll get:

$$0.25q + 0.10\boldsymbol{d} = 3.30$$

$$0.25q + 0.10(\boldsymbol{3q}) = 3.30$$

$$0.25q + 0.30q = 3.30$$

$$0.55q = 3.30$$

$$q = \frac{3.30}{0.55} = \frac{330}{55} = \frac{66}{11} = \boldsymbol{6}$$

If **q = 6**, then because $d = 3q$, we know that **d = 3 · 6 = 18**, right? And does it work? The money from 6 quarters would be **6**($0.25) = $1.50, and the money from 18 dimes would be **18**($0.10) = $1.80, and those add up to $3.30. Yep!

Answer: 6 quarters and 18 dimes

QUICK NOTE   In most of these mixture problems, our goal will be to create two true statements from the information we're given: a "thing" equation (in the above example, it was $d = 3q$) and a "money" equation (in the above example, it was $0.25q + 0.10d = 3.30$). <u>One equates things to things, and the other equates dollars to dollars.</u>

• • • • • • • • • •

* If that went too fast, check out pp. 93–94 in *Kiss My Math*.
[†] See Chapter 12 to review solving systems of equations with substitution.

Back to the Almonds

Warming up to the party-mix problem: It makes sense that if almonds cost \$5.50/lb and we bought 2 pounds of them, then we'd be spending (\$5.50)(2) = \$11, right? Well, if the number of pounds of almonds is n, then we'd use n instead of 2, so we'd be spending (\$5.50)(n) = **\$5.50n**. Make sense? Okay, let's see that problem again.

If almonds cost \$5.50/lb and cranberries cost \$7.50/lb, and we want 3 lbs of this mixture to cost \$20, then how much of each ingredient should we buy?

We'll label n = pounds of almonds and c = pounds of cranberries. Since we know the total weight needs to be 3 pounds, our "thing" equation is **$n + c = 3$**. How about our "money" equation? We know the *cost* of the almonds can be written as \$5.50n, and the *cost* of the cranberries is \$7.50c. We know the <u>total cost</u> will be \$20, so we can write:

$$\underbrace{5.50n}_{\substack{\text{cost from} \\ \text{almonds}}} + \underbrace{7.50c}_{\substack{\text{cost from} \\ \text{cranberries}}} = \underbrace{20}_{\text{total cost}}$$

In our "thing" equation, $n + c = 3$, we can solve for c, and then substitute $c = \mathbf{3 - n}$:

$$5.50n + 7.50\mathbf{c} = 20$$

$$5.5n + 7.5(\mathbf{3 - n}) = 20$$

$$5.5n + 22.5 - 7.5n = 20$$

$$-2n = -2.5 \quad \text{Subtracted 22.5 from both sides and combined like terms}$$

$$n = \frac{-2.5}{-2} = \frac{2.5}{2} = \frac{25}{20} = \frac{5}{4}$$

So, $\mathbf{n = \dfrac{5}{4}}$.

We know that $c = 3 - n$, so $c = 3 - \dfrac{5}{4} = \dfrac{12}{4} - \dfrac{5}{4} = \dfrac{7}{4} = \mathbf{1\dfrac{3}{4}}$ **lbs**.

And $n = \dfrac{5}{4} = \mathbf{1\dfrac{1}{4}}$ **lbs**. It's never a bad idea to brush up on fraction skills!

Answer: We should buy $1\dfrac{1}{4}$ lbs of almonds and $1\dfrac{3}{4}$ lbs of cranberries. Yum.

Step By Step

Mixture problems . . . involving money

Step 1. Look at your *units*. There will be "things" (like coins, pounds, etc.), "money" (like dollars or cents), and *"cost per thing"* units: This could be dollars/ounce, cents/coin, dollars/pound, dollars/ticket, or something else! Take the opportunity to make sure your units are consistent, and make any necessary unit conversions.

Step 2. Label the unknowns, and set up the equations you can—typically, a "thing" equation between *numbers of things* and a "money" equation between *total costs*.

Step 3. Use variable substitution to create one equation with one unknown.

Step 4. Solve for the variable and for any other variables necessary for the answer.

Step 5. Reality check—does the answer make sense? If so, we're done!

And... Action! Step By Step In Action

You've organized a live dance show to raise money for charity. You're charging $8 for student tickets and $12 for adult tickets. If you sold three times as many student tickets as adult tickets and you made a total of $540, how many of each type of ticket did you sell?

Step 1. First things first: What are our units? Hmm, the "things" are the student tickets and adult tickets, right? And our money unit is dollars.

Step 2. We'll label the number of student tickets *s* and adult tickets *a*. We sold three times as many student tickets as adult tickets, so our "thing" equation is **s = 3a**.

Student tickets cost $8 each, so if we sold *s* of them, we'd make **8s** dollars. Similarly, the money made from the adult tickets is **12a**. Since the money from student tickets and adult tickets totals $540, then we can create this totally true statement, equating dollars to dollars:

$$8s + 12a = 540$$

money from student tix   money from adult tix   TOTAL money from all tix

Step 3. We'll use **s = 3a** to substitute in the above equation: 8(**3a**) + 12a = 540.

Step 4. Solving it, we'll simplify and then isolate *a*:

$$8(3a) + 12a = 540$$

$$24a + 12a = 540$$

$$36a = 540$$

$$a = \frac{540}{36} = \frac{60}{4} = 15$$

So, **a = 15**. Now we can find *s*, since *s* = **3a** = 3(**15**) = 45. So, **a = 15** and **s = 45**.

Step 5. Does the answer make sense? We can plug our answers into the money equation from the previous page, like this: $8(**45**) + $12(**15**) = $360 + $180 = $540. Yep!

Answer: 45 student tickets and 15 adult tickets were sold.

Take Two: Another Example

You just made a beautiful beaded necklace using 3 kinds of beads: clear, blue, and green—like the ocean. Now your friend wants to know how much the green beads cost so she can get some, but you can't remember! You do remember that the clear beads cost twice as much as the green beads, and that the green beads cost three times as much as the blue beads. If you bought 8 of each bead, and according to your receipt, the whole thing cost $2.40, can you tell her how much the green beads cost?

Steps 1 and 2. This time we are given different information. We know the numbers of objects this time but not their individual costs. That's okay—let's see what happens. The "things" are the beads, and the cost is in dollars. Let's assign variables, and to keep from getting confused, let's use capital letters for the *costs* of each type of bead.

Let's label: *C* = cost for 1 clear bead, *B* = cost for 1 blue bead, and *G* = cost for 1 green bead. So **8C** = cost of 8 clear, **8B** = cost for 8 blue, and **8G** = cost for 8 green.

Step 3. Let's do the money equation first: **8C + 8B + 8G = 2.40**. This represents the total cost for all 3 types of beads. With me so far? Now we need our "thing" equation. But wait, we already know how many of each bead we have: 8 of each type. Hmm. What else do we know? We know the clear beads cost twice as much as the green beads, so $C = 2G$, and that the green beads cost three times as much as the blue, so $G = 3B$.

Step 4. We have 3 equations and 3 unknowns, so we can solve it. Notice that we can write all three bead costs in terms of B. After all, **$G = 3B$**, and since $C = 2G$, we substitute $G = 3B$ into it and get: $C = 2(3B) = 6B \rightarrow$ **$C = 6B$**. This means we can rewrite our big equation and solve for B:

$$8C + 8B + 8G = 2.40$$
$$8(\mathbf{6B}) + 8B + 8(\mathbf{3B}) = 2.40$$
$$48B + 8B + 24B = 2.4$$
$$80B = 2.4$$
$$B = \frac{2.4}{80} = \frac{24}{800} = \frac{3}{100}$$

What does it mean that $B = \frac{3}{100}$ dollars? Well, $\frac{3}{100} = $ **0.03** dollars = **3 cents**, darlin'! Next, we can solve for G, which is what the question asked for: $G = 3B = 3(\mathbf{0.03}) = $ **\$0.09**.

Step 5. Does this make sense? Let's see, $C = 6B = 6(3) = 18$ cents, so we could plug our answers back in and find that it checks out perfectly: $8(0.18) + 8(0.09) + 8(0.03) = 2.4 = $ **\$2.40**. Yep!

Answer: The green beads cost \$0.09 each; in other words, 9¢ each.

That problem didn't have a "thing" equation. Instead, it gave us extra relationships between costs. It's good to be flexible! Just write down any true math statements you can, using the relationships given, and the rest will fall into place. That's good advice for *all* word problems, actually!

 QUICK NOTE Instead of money, some problems will involve *points* like the game problem below. They work the exact same; after all, we're giving "things" a certain value, whether in dollars or points, and they can sometimes be negative. Check out problem 1 and problem 6 on the following pages.

Doing the Math

Solve these mixture problems. I'll do the first one for you.

1. Kayo just played a new game called "Sing Like a Rock Star." She gets 10 points for each correct note, but 15 points get deducted for each wrong note. She sang a total of 220 notes and got a score of 1700 points. How many notes did she sing correctly?*

<u>Working out the solution</u>: Okay, so our "things" are notes, and the "money" is points. "Cost per thing" will be 10 points per correct note and −15 points per wrong note. Let's call the number of correct notes c and the number of wrong notes w. Then she earned $10c$ points, but she got $15w$ points deducted, right? This means **$10c + (-15)w = 1700$**. And we also know that the *total* number of notes is 220, so $c + w = 220$; in other words, $c = \mathbf{220 - w}$, which we can substitute into the bigger equation:

$$10c + (-15)w = 1700$$

$$10(\mathbf{220 - w}) + (-15w) = 1700$$

$$2200 - 10w - 15w = 1700$$

$$-25w = -500$$

$$w = \frac{-500}{-25} = \frac{500}{25} = \frac{100}{5} = \mathbf{20}$$

Because $w = 20$, that means she sang 20 wrong notes. And because $c + w = 220$, that means she sang 200 notes correctly. Hey, that's pretty darn good! Let's plug in the answer to make sure we didn't make any careless mistakes:
$10c + (-15)w \stackrel{?}{=} 1700 \rightarrow 10(\mathbf{200}) + (-15)(\mathbf{20}) \stackrel{?}{=} 1700$
$\rightarrow 2000 - 300 \stackrel{?}{=} 1700$. Yep, we're good!

Answer: Kayo sang 200 notes correctly.

. . . .

* If it makes it easier to follow the solution, replace the word *points* with *cents*. Then it will feel more like the money problems we've been doing.

2. Olivia has 10 coins—all dimes and nickels. If she has a total of 80¢, how many nickels does she have?

3. Debbie sells two kinds of shoes at her store—high heels and flats. The high heels cost $80 each, and the flats cost $60 each. Today she sold a total of 20 pairs of shoes and made $1,440 in sales. How many pairs of high heels did she sell?

4. Max bought a 15-pound mixture of gumballs and jelly beans, which cost a total of $9. If gumballs cost 30¢/lb and jelly beans cost $1.20/lb, then how many pounds of jelly beans did he buy?

5. Cheyenne has 5 more dimes than quarters, half as many nickels as quarters, and no pennies. If she has a total of $2.00, how many dimes does she have?

6. To earn her allowance, Taylor does chores around the house. She makes $2.00 for each chore she does and gets $1.50 *deducted* for each chore she forgets to do. There's a total of 13 possible chores each week. This week, she made $19. How many chores did she forget to do? *(Hint: This is similar to problem 1.)*

(Answers on p. 409)

What's the Deal?

Just like with rate of work and motion problems, the units are actually canceling in these mixture problems. See, because the "cost per thing" fraction has a *unit* on the bottom, it cancels with that *same unit* outside the fraction, and we end up with cost, alone:

$$\frac{5.5 \text{ dollars}}{1 \text{ lb}} \times n \text{ lbs} = 5.50n \text{ dollars}$$

It's worth noticing, that's all. For more on units canceling, see pp. 206–7.

Liquids: Percents of a Special Ingredient!

Maybe someone in your family is known for his or her famous chili or cookies containing a secret special ingredient, passed down only by word of mouth from generation to generation. No? Well, me neither. But this next type of mixture problem involves tracking a *special ingredient* inside a liquid. Keep that in mind, and you'll be golden!

If I have a big container of chocolate milk that is 20% chocolate syrup (and the rest is milk or whatever), and you have a big container of chocolate milk that is 50% chocolate syrup, how much of each should we use to create an 8-oz glass of chocolate drink that consists of 40% chocolate syrup?

So if Liquid A contains 20% chocolate syrup, then 100 oz of Liquid A has **20 oz** of *pure* chocolate syrup in it, right? That's because 20% of 100 oz is **20 oz**; in other words, $(0.20)(100) = 20$. (And milk or whatever else makes up the other 80 oz.) Similarly, if we had 30 oz of Liquid A, then the amount of pure chocolate syrup would be 20% of 30 oz = $(0.20)(30) = 6$ **oz**, and the other 24 oz would be milk or whatever. So if we label *a* as the number of total ounces of Liquid A, then here's the amount of pure chocolate syrup contained in those *a* ounces: **0.20a**. Make sense? Similarly, if we have *b* ounces of Liquid B, then the amount of actual pure chocolate syrup inside would be **0.50b**.

Once we mix them together, we're supposed to end up with 8 oz of a 40% chocolate syrup liquid, which means our final mixture will have

$(0.40)(8) = $ **3.2 oz** of <u>pure</u> chocolate syrup in it. Let's write a true statement equating amounts of pure chocolate syrup:

$$0.20a + 0.50b = 3.2$$

<u>⌣</u> Ounces of <u>pure</u> syrup Contributed from Liquid A   <u>⌣</u> ounces of pure syrup from Liquid B   <u>⌣</u> ToTAL ounces of <u>pure</u> syrup in mixture, Liquid C

Notice that on either side of the equation, we are equating the ounces of *pure chocolate syrup* only, not the ounces of total liquid. Speaking of the total liquid, we also know that the total liquid from Liquid A and Liquid B must equal 8 ounces, which means $a + b = 8$; in other words, **b = 8 – a**. Let's substitute **8 – a** for **b** into the previous equation (dropping any unnecessary zeros) and solve it:

$$0.2a + 0.5\boldsymbol{b} = 3.2$$

$$0.2a + 0.5(\boldsymbol{8 - a}) = 3.2$$

$$0.2a + 4 - 0.5a = 3.2$$

$$-0.3a = -0.8 \quad \text{subtracted 4 from both sides and combined like terms}$$

$$a = \frac{-0.8}{-0.3} = \frac{8}{3} = 2\frac{2}{3} \text{ oz}$$

which means that $b = 8 - a = 8 - 2\frac{2}{3} = 5\frac{1}{3}$ **oz.**

So to make 8 ounces of chocolate milk containing 40% chocolate syrup, we'd combine $2\frac{2}{3}$ **oz of Liquid A** (which is 20% syrup) and $5\frac{1}{3}$ **oz of Liquid B** (which is 50% syrup). Yum!

By the way, it can be easier to first write the equation with *percents* and then just convert the percents to decimals. So we could have started the problem like this:

$$(20\%)a + (50\%)b = (40\%)(8)$$

$$0.20a + 0.50b = 0.40(8)$$

The most important thing is to understand that we are creating a true statement, equating amounts of *pure chocolate syrup* on both sides (or whatever the special ingredient is!).

Reality Math

Dress Dye Dilemma

*O*kay, so your favorite white dress got a stain that won't fully come out. Bummer. But you've decided to make the best of the situation: You're going to hide the stain by dying the entire dress an antique yellow—beautiful! (Antique yellow can be made from diluting brown dye.) You have 8 cups of a dye solution that is 30% brown dye, but you read online that 5% is best for that antique look. Otherwise, the dress might come out sort of tan. Can you figure out how many cups of water to add to your existing dye solution to get the perfect dye color for your dress? You bet! Check out p. 235 to see how.

Step By Step

Mixture problems involving liquids and special ingredients:

Step 1. Identify the special *ingredient* inside the liquids that's being asked about: Is it chocolate? Salt? Alcohol? Antifreeze? Then imagine the containers of liquids. Label them Liquid A and Liquid B, and the mixture container will be Liquid C.

Step 2. Label the unknowns, and create a couple of true statements based on the information given. Typically, one equation will be between the *total amounts* of Liquids A, B, and C. The other will usually look something like this:

(% in Liq. A)(oz of Liq. A) + (% in Liq. B)(oz of Liq. B) = (% in Liq. C)(oz of Liq. C)

. . . which describes the total amounts of *special ingredient* found in each container of liquid. Be sure to convert the percents to decimals.

Step 3. Use substitution to create one equation in one variable, and solve it.

Step 4. If you haven't yet, make sure you have answered what the problem asked. Done!

With so many decimals in these problems, it can be easy to make a mistake! Try multiplying both sides of the equation by 100 or 1000—whatever will make all the decimal places go away. Just be sure to do the same thing to each entire side of the equation, count your decimal places, and remember to distribute.*

Remember the dress dye dilemma from p. 234? Here's how to handle it:

And... Action! Step By Step In Action!

How much water should we add to 8 cups of a 30% brown dye solution so that it is only 5% brown dye and will make our dress look antique?

Step 1. *Brown dye* is the special ingredient. In this case, Liquid A is 30% brown dye, and Liquid B is water, which is 0% brown dye, right? Our mixture, Liquid C, will be 5% brown dye.

Step 2. Our unknowns are the *amount* of Liquid B, which we'll call b, and also the *amount* of Liquid C we will end up with. Luckily, there's a built-in relationship here: We have 8 cups of liquid that we're adding b cups to, so that means we'll end up with $8 + b$ total cups of Liquid C. Make sense? And we're ready to set up our equation:

$$(30\%)(8) + (0\%)b = 5\%(8 + b)$$

$$0.30(8) + 0(b) = 0.05(8 + b)$$

Step 3. Let's solve for b! Pretty crazy that the 0 will cancel out that middle term. But this makes sense when you think about it: Because we're not adding any brown dye to the mixture, *the actual amount of brown dye in*

. . . .

* Seriously, it's so easy to forget to distribute.

Liquid A is indeed equal to the *actual amount of brown dye* in Liquid C, and that's what this equation says. This stuff really blows my mind sometimes. Okay, let's find the b that makes this statement true!

$$0.30(8) + 0(b) = 0.05(8 + b)$$

$$2.4 = 0.4 + 0.05b$$

I don't like all these decimals; let's multiply both sides by 100:

$$\mathbf{100}(2.4) = \mathbf{100}(0.4 + 0.05b)$$

$$240 = 40 + 5b$$

$$200 = 5b$$

$$b = \frac{200}{5} = \mathbf{40}$$

Step 4. So **b = 40**, which means we'd need to add 40 cups of water in order to end up with a final solution that is only 5% brown dye for a beautiful, antique-looking dress.

Answer: We'd need to add 40 cups of water.

We'd end up with 8 + 40 = 48 cups of solution, which is 3 gallons of dye solution.* So we could dye a bunch of other clothes, too!

QUICK NOTE   Comparing Types of Mixture Problems: Money vs. Liquid

Notice that, just like in money mixture problems from earlier in this chapter, *liquid* mixture problems will often involve a "thing" equation, too. This will be a true statement equating the ounces or milliliters (or cups or gallons, etc.) of the <u>total liquids</u> we have.

The bigger equation in money mixture problems is the "money" equation, equating dollars on both sides. By contrast, the bigger equation in liquid problems equates the ounces or milliliters (or cups or gallons, etc.) of <u>just the special ingredient</u> on both sides.

. . . .

* $\dfrac{1 \text{ gallon}}{16 \text{ cups}} \times 48 \text{ cups} = \dfrac{48}{16}$ gallons = **3 gallons**

Depending on the problem, we won't always need both equations. (See problem 1 below.) Just focus on creating true statements about the liquids and see where it takes you.

Doing the Math

Solve these liquid mixture problems. I'll do the first one for you.

1. Your adorable little sister was playing with her mermaid dolls last night. She filled your bathtub with water and then added salt to make 40 gallons of salt water; in fact, it's 4% salt, almost like the ocean! Later, she forgot to drain the tub. Now, the next morning, it's 5% salt. How much water must have evaporated?

<u>Working out the solution</u>: Let's not panic. We'll call w the amount of water that evaporated. So the gallons of liquid in the tub after evaporation would be $40 - w$, right? Our special ingredient is pure salt, and notice that the amount of pure salt before and after evaporation will be the same: That's the key to this problem. Before evaporation, the amount of pure salt is $(4\%)(40)$, right? That's $(0.04)(40) = $ **1.6** gallons of pure salt. After evaporation (in the tub, which now is 5% salt), the amount of pure salt can be expressed as $(5\%)(40 - w) = $ **$(0.05)(40 - w)$**. Make sense? Because the total amount of pure salt in the tub before and after evaporation is the same, we can set them equal to each other, and we get **$1.6 = (0.05)(40 - w)$** $\rightarrow$ $1.6 = 2 - 0.05w$ $\rightarrow$ $-0.4 = -0.05w$ (multiplying both sides by -100) $\rightarrow$ $40 = 5w$ $\rightarrow$ $8 = w$. So **8 gallons** ~~of water~~ must have evaporated. Wacky!

~~Answer:~~ **8 gallons of water evaporated.**

2. You're throwing a luau! Your favorite guava-pineapple juice is 30% guava juice, but the store ran out. Instead, they have one that is 40% guava juice (Brand A) and one that is 10% guava juice (Brand B). How much of each should you use to create 2 gallons of a mixture that is 30% guava juice?

3. Samantha has a jug of 4 gallons of glitter paint, which contains 10% glitter. How much pure glitter should she add so the jug becomes 70% glitter? *(Hint: Pure glitter is just a "liquid" that is 100% pure glitter, and 100% is 1 as a decimal.)*

4. If Chipmunk Charlie's Nut Mix contains 25% hazelnuts and Rabbit Rosy's Nut Mix contains 45% hazelnuts, how many ounces of each will make 20 oz of a nut mix containing 40% hazelnuts? *(Hint: Since we're given percents, it works the same way as liquids.)*

5. This time, your sister made a 6% salt solution in your tub. Overnight, 5 gallons of water evaporated, and in the morning, the liquid was 8% salt. How much saltwater did the tub start out with the night before?

6. 14-carat gold is about 60% pure gold, and 18-carat gold is 75% pure gold. How much pure gold should be melted and added to 40 grams of melted 14-carat gold to make 18-carat gold? *(Hint: Don't be distracted by the 14 and 18—because you won't need these numbers!)*

(Answers on p. 409)

Takeaway Tips

Most mixture problems involve two separate quantities to consider: objects & their values, or total liquids & their special ingredients. Label whatever isn't given.

For most mixture problems involving money, one equation will be between the amounts of "things" and the other will equate dollars to dollars (or cents to cents).

For liquids problems, remember that the amount of pure special ingredient in the first two liquids will combine to equal the amount of *pure* special ingredient in the mixture; that will be your main equation. The other equation, if you need it, will be between the amounts of total liquids.

Chapter 17

Ms. Exponent Gets Her Whip

The Product, Power, and Quotient Rules for Exponents

$\mathscr{P}$erhaps you recall Ms. Exponent from *Kiss My Math*. She's the high-powered executive who runs her business from up high in her corner penthouse office and orders mergers and martinis with a quick call to her cute assistant named Brett. Yes, I had a feeling you'd remember her.

Well she's back, and in this book, Ms. Exponent gets out her whip and means business. The best part? You shall soon have at *your* command the power she wields to make numbers and variables do as she pleases with a mere flick of the wrist and a crack of the whip. (You know, tips to make your homework easier.)

First, let's brush up on the basics.

Mini Review of
Intro to Exponents, as Taught in Kiss My Math

Ring Ring *What's It Called?*

An **exponent** is a powerful little number in an exponential expression that indicates <u>how many times</u> a **base** gets multiplied times itself. That's right, Ms. Exponent is the one giving the orders. For example, in the expression 3^4, the *exponent* is 4, and the *base* is 3, so the 3 should multiply times itself 4 times: $3^4 = 3 \times 3 \times 3 \times 3 = 81$.

Pay careful attention to the placement of parentheses, and remember that exponents affect the thing they are actually *touching*. If an exponent touches the outside of parentheses, then the exponent affects everything inside, including negative signs. For example:

$$(-3)^4 = 81 \text{ but } (-3^4) = -3^4 = -81$$

Also, remember that you can <u>distribute exponents</u> inside a set of parentheses when the inside is a *product of factors**; in other words, when the *only* operations inside the parentheses are multiplication and division[†]—NOT addition or subtraction. Put another way: For bases a and b and exponent m, we can distribute the exponent over multiplication and also over division:

$$(a \cdot b)^m = a^m \cdot b^m$$

$$\left(\frac{a}{b}\right)^m = \frac{a^m}{b^m}$$

So $9\left(\frac{2}{3}\right)^2 = 9\left(\frac{2^2}{3^2}\right) = 9\left(\frac{4}{9}\right) = \mathbf{4}$.

But we cannot distribute exponents over addition or subtraction, so for example,

$$(a + b)^m \neq a^m + b^m$$

---

* See p. 16 for the definition of *product of factors*.
[†] Division in these kinds of expressions is almost always seen in fraction notation.

This was a super-quick overview of the basics of exponents. If this hasn't been totally review so far, see Chapters 15 and 16 in *Kiss My Math*.

Ms. Exponent's Other Talents

Oh yes, she has many.

What if you wanted to simplify $3^2 \cdot 3^4$ or $(x^3)^7$ or even $(4xy)^3x^4yz$? You could expand all the multiplication out, but that would take forever—and you just *know* that Ms. Exponent has thought of an easier way. I'll show you her latest tricks in just a moment. First, take a look:

$$2 \times 2 \times 2 \times 2 \times 2 \times 2 \times 2 = 2^7$$

Since there are seven 2's, we know we can write it as 2^7, which, when multiplied out, equals 128. But let's rewrite that long string of 2's by grouping them like this, and then seeing how we'd write them with exponents:

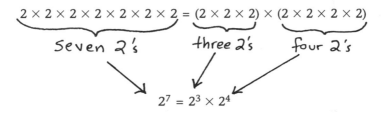

Notice the exponents on the right side: 3 + 4 = **7**. And this makes perfect sense, because all we're doing is *counting* how many 2's are multiplying on both sides, right? I feel a shortcut alert coming . . .

Shortcut Alert

Product Rule for Exponents: AKA, the "Add the Exponents" Shortcut

To multiply two exponential expressions with the *same base*, just keep the base the same and <u>add</u> their exponents! So:

If the base *b* is a real number and the exponents *m* and *n* are positive integers, then:

$$b^m \cdot b^n = b^{m+n}$$

For example:

$$3^5 \cdot 3^2 \ = \ 3^{5+2} \quad = \ 3^7$$

$$x^3 \cdot x^7 \ = \ x^{3+7} \quad = \ x^{10}$$

$$(-y)^2 \cdot (-y)^3 \ = \ (-y)^{2+3} \ = \ (-y)^5$$

I know all the variables and exponents can look a little daunting, but read the above examples slowly and notice that to *multiply* the expressions, we just *add up* the exponents to get the answer. See? It's not so bad! This is a shining example of how Ms. Exponent can accomplish so much by doing *so* little. (SFX: Whip crack!)*

Watch Out!

Always make sure that *the bases are the same* before adding exponents. If you see something like $3^3 \cdot 6^2$ or $2^5 \cdot y^6$, unfortunately, there's no shortcut for combining them—because they have different bases. Your only option is to multiply everything out. (You'd get 972 and $32y^6$, respectively, by the way.)

Let's multiply $\left(-\frac{1}{2}\right)^3 \cdot \left(-\frac{1}{2}\right)^3$.

Don't let the negative signs get you down. This isn't so bad when we take things a step at a time! Because the base, $\left(-\frac{1}{2}\right)$, is the same, we can totally ignore *what* the base is; the $\left(-\frac{1}{2}\right)$ might as well be a flower or an umbrella or something. We simply add the exponents:

$$(❀)^3 \cdot (❀)^3 \ = \ (❀)^{3+3} \ = \ (❀)^6$$

$$\left(-\frac{1}{2}\right)^3 \cdot \left(-\frac{1}{2}\right)^3 \ = \ \left(-\frac{1}{2}\right)^{3+3} \ = \ \left(-\frac{1}{2}\right)^6$$

. . . .

* SFX stands for "sound effects." Get it? Sound "F X?" SFX is printed in movie and TV scripts all the time: "Abby walked down the sidewalk, daydreaming about math again. SFX: RING! Startled, she looks at her cell phone . . . and it's him!" (And here endeth the lesson on screenwriting.)

Fabulous. Okay, *now* let's look at the details inside the parentheses. Since the number inside is negative, but the exponent on the outside is *even*, we know that the negative signs will cancel each other. We can just get rid of the negative sign and distribute the exponent to the top and bottom of the fraction:

$$\left(-\frac{1}{2}\right)^6 = \left(\frac{1}{2}\right)^6 = \frac{1^6}{2^6} = \frac{1}{64}*$$

Answer: $\frac{1}{64}$

Watch Out!

We are allowed to use the "add the exponents" shortcut only when we're *multiplying* expressions, not adding or subtracting. So, if we have $x^3 + x^2$, there's *no way* to combine them or to simplify this in any way (unless somebody tells us what the value of x is, of course).

Also notice: $2x^3 + 4x^3 = 6x^3$, not $6x^6$. This is just adding like terms,[†] after all. We'd have to *multiply* them together to add their exponents: $2x^3 \cdot 4x^3 = 8(x \cdot x \cdot x)(x \cdot x \cdot x) = 8x^6$. From time to time, it's a good idea to multiply it out like this, just so you remember *why* we're allowed to add exponents during multiplication. It'll keep you in great shape!

QUICK NOTE   EASY FORM: Whenever we see a variable that doesn't seem to have an exponent, its exponent is actually **1** (NOT zero). It's just invisible, that's all. I like to write things out in what I call "easy form" with dot multiplication and invisible exponents written in, which makes it easier to see what's going on:

$$x(x^2) \rightarrow x^1 \cdot x^2 = x^{1+2} = x^3$$

. . . .

* Or we could have done $\left(-\frac{1}{2}\right)^6 = \frac{(-1)^6}{2^6} = \frac{1}{64}$. For more on negative signs with exponents, see Chapter 15 in *Kiss My Math*.

[†] To review combining like terms, see Chapter 9 in *Kiss My Math*.

$$(x^2y) \cdot (x^5y^8) \cdot x^3$$

"easy form":

$$= \; x^2 \cdot y^1 \cdot x^5 \cdot y^8 \cdot x^3$$

Then let's rearrange the factors so the x's are together and the y's are together, and finish by adding the exponents on similar bases:

$$= x^2 \cdot x^5 \cdot x^3 \cdot y^1 \cdot y^8$$
$$= x^{2+5+3} \cdot y^{1+8} \; = \; x^{10}y^9$$

I mean, it's rearranging and adding, how hard can it be? I know . . . famous last words.

Step By Step

Multiplying exponential expressions with the same bases, using the Product rule:

Step 1. Distribute exponents over multiplication/division in parentheses, if applicable.

Step 2. Rewrite the expression in "easy form": Use dot multiplication so everything's nice and spread out, and write in any invisible 1 exponents.

Step 3. Bring the coefficient(s) to the front, group the bases together, and for each base, *add* up their exponents. Done!

And... Action! Step By Step In Action

Let's simplify $a^9(-2b)^2 \cdot ab^5$.

Step 1. First, let's deal with $(-2b)^2$. We could distribute the exponent, $(-2)^2(b)^2 = \textbf{4}b^2$, or we could just multiply it out: $(-2b)^2 = (-2b)(-2b) = \textbf{4}b^2$. Either way is fine! Now our full problem looks like this: $a^9(4b^2) \cdot ab^5$

Step 2. Time to rewrite it in easy form: $a^9 \cdot 4 \cdot b^2 \cdot a^1 \cdot b^5$

Step 3. Bringing the 4 to the front and collecting bases, we get $4 \cdot a^9 \cdot a^1 \cdot b^2 \cdot b^5 = 4a^{9+1} \cdot b^{2+5} = \textbf{4}\boldsymbol{a}^{10}\boldsymbol{b}^7$. Ah, much nicer looking.

Answer: $4a^{10}b^7$

244 EXPONENTS AND SQUARE ROOTS

Doing the Math

Multiply these expressions using the Product rule. I'll do the first one for you.

1. $(4xy)^3 \cdot x^4 \cdot y \cdot z^2$

<u>Working out the solution</u>: Before we do anything else, let's distribute that 3 over the multiplication inside the parentheses: $(4xy)^3 = 4^3 \cdot x^3 \cdot y^3$. Knowing that $4^3 = 64$ and writing it in easy form, the <u>full</u> expression looks like this: $64 \cdot x^3 \cdot y^3 \cdot x^4 \cdot y^1 \cdot z^2$, which we can rearrange to collect similar bases together: $64 \cdot x^3 \cdot x^4 \cdot y^3 \cdot y^1 \cdot z^2$ (adding exponents) $= 64 \cdot x^{3+4} \cdot y^{3+1} \cdot z^2 = 64 \cdot x^7 \cdot y^4 \cdot z^2$.

Answer: $64x^7y^4z^2$

2. $x \cdot y \cdot x^5 \cdot y^6$

3. $y \cdot (5y)^2 \cdot y^3$

4. $(-2x)^2 \cdot xyz$

(Answers on p. 409)

Women Supporting Each Other!

Go out of your way to let a woman know how much you admire her. If you've been quietly thinking how strong or confident your mom, aunt, or teacher is, let her know! Women love to feel appreciated. And for girls younger than you, go out of your way to *help* them, whether by tutoring or giving some good advice. They look up to you and want to be like you! Guys have always had their "boys' clubs," and it's time we young women learned to really support each other, don't you agree?

The Power Rule

I bet you thought Ms. Exponent was powerful when she added exponents, but you ain't seen nothin' yet. Consider these:

$$(4^2)^3 \qquad (x^3)^4 \qquad (3xy^2)^4$$

Hmm, raising powers to more powers—wacky! Okay, let's look at $(4^2)^3$ and pretend that 4^2 is a flower or something (just so we don't concern ourselves with the 2 exponent). Let's multiply it out to see what's going on: $(4^2)^3 = (4^2)(4^2)(4^2) = 4^2 \cdot 4^2 \cdot 4^2$. Okay, now let's add exponents: $4^{2+2+2} = 4^6$. So, $(4^2)^3 = 4^6$. Interesting, it seems like we could just *multiply* those exponents together to get the same answer: $(4^2)^3 = 4^{2 \cdot 3} = 4^6$.

Let's keep investigating. How about something like $(x^3)^4$? Well, let's multiply it out: $(x^3)^4 = x^3 \cdot x^3 \cdot x^3 \cdot x^3 = x^{3+3+3+3} = x^{12}$. And we've just discovered that $(x^3)^4 = x^{12}$. Again, we could have multiplied those exponents to get our answer, because $3 \times 4 = 12$.

Amazing. Ms. Exponent is so powerful that she can raise a power to another power with just a little multiplication! Now *that's* power. In fact, that's the Power *rule*:

Shortcut Alert

Power Rule for Exponents: AKA, the "Multiply the Exponents" Shortcut

If the exponents m and n are non-negative integers,* and the base b is a real number, then:

$$(b^m)^n = b^{m \cdot n}$$

For example, $(x^5)^4 = x^{20}$.

---

* In the next chapter, we'll see that m and n can also be negative as long as b isn't zero. (This is also true for the Product rule.)

Step By Step:

Using the Power rule: AKA, the "multiply the exponents" shortcut:

Step 1. When there is an exponent on the outside of parentheses with more exponents *inside,* before using the Power rule to multiply the exponents, make sure the only operations inside the parentheses are multiplication/division,* *not* addition/subtraction.

Step 2. Write in 1 for any invisible exponents. It makes things easier.

Step 3. Distribute the exponent to *each base* inside the parentheses, and multiply the exponent times the exponents that are already inside.

And... Action! Step By Step In Action

Let's simplify $(3xy^2)^4$.

Step 1. No addition or subtraction is inside the parentheses—fantastic.

Steps 2 and 3. We'll write in a 1 for *x*'s and 3's exponents. Then, we'll distribute the 4 to the 3, *x*, and y^2, and we'll use our shortcut to multiply exponents:

$$(3^1 x^1 y^2)^4 \;=\; 3^{1\cdot4} x^{1\cdot4} y^{2\cdot4} \;=\; \mathbf{81x^4y^8}$$

And that's our answer.

Watch Out!

As I mentioned in the Step By Step, you can only use the "multiply the exponents" shortcut when whatever's inside the parentheses is stuck together with multiplication and division—not addition or subtraction! So, if you have something like $(2y^3 + 5)^2$, you might be tempted, but *do not* distribute that 2 exponent inside the parentheses. To be clear, $4y^6 + 25$ is <u>not</u> the answer. There's addition inside, and that changes everything.

• • • • • • • • • •

* Division is usually in fraction notation.

When in doubt, multiply it out! In this case, multiplying it out shows us that $(2y^3 + 5)^2 = (2y^3 + 5)(2y^3 + 5)$. Make sense?*

QUICK NOTE   Despite the WATCH OUT on the previous page, we *can* use the Power rule to multiply 3×2 on the expression below. Do you see why? Pay close attention to the *location* of the exponents, which is a little different this time.

$$[(a + 5b)^3]^2$$

Addition is technically part of this expression, but since we're not attempting to distribute the exponent *over* the addition (which would be bad!), we're totally fine. Check it out.

$$[(a + 5b)^3]^2 = (a + 5b)^6$$

Think of it this way: We could have mentally substituted $(a + 5b)$ for another variable, like w. Then it's easy to see how this was totally allowed, since $[(w)^3]^2 = (w)^6$.

How about one with division (in fraction form)?

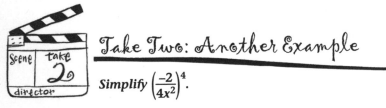

Take Two: Another Example

Simplify $\left(\dfrac{-2}{4x^2}\right)^4$.

Steps 1 and 2. With no addition or subtraction going on here, it's safe to distribute the exponent to the insides.

Step 3. Distributing the exponents, we get $\dfrac{(-2)^4}{(4x^2)^4} = \dfrac{(-2)^4}{4^4 x^{2 \cdot 4}} = \dfrac{16}{256x^8}$.

We can reduce this (because $256 = 16 \cdot 16$) and get $\dfrac{16}{256x^8} = \dfrac{1}{16x^8}$. Voilà!

Answer: $\dfrac{1}{16x^8}$

. . . .

* In case you're curious, $(2y^3 + 5)(2y^3 + 5) = 4y^6 + 20y^3 + 25$. Don't worry though, it's not as complicated as it looks, especially when you're at the hair salon. (You'll know what I mean when we learn to do this with FOIL in Chapter 21!)

QUICK NOTE   With the Power rule, it doesn't matter *which* exponent is inside and *which* is outside. Either setup will give the same answer, because the exponents multiply either way, right? For example, $(x^3)^5 = (x^5)^3$; they both equal x^{15}.

Shortcuts for Exponents: Keeping 'Em Straight:

These two rules in particular can be hard to keep straight because they both involve multiplication, so it's helpful to see them side by side.

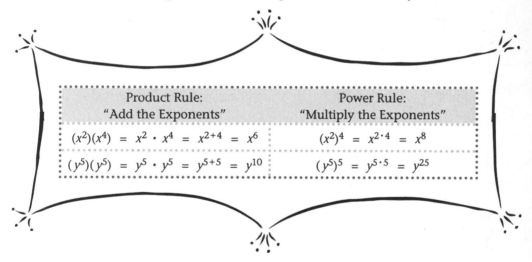

Product Rule: "Add the Exponents"	Power Rule: "Multiply the Exponents"
$(x^2)(x^4) = x^2 \cdot x^4 = x^{2+4} = x^6$	$(x^2)^4 = x^{2 \cdot 4} = x^8$
$(y^5)(y^5) = y^5 \cdot y^5 = y^{5+5} = y^{10}$	$(y^5)^5 = y^{5 \cdot 5} = y^{25}$

Remember: If you ever feel confused about these two shortcuts, you can always multiply out an expression to remind yourself what's really going on and *why* the shortcuts work.

Watch Out!

With expressions like problem 6 on the next page, $(xy^2)(y^3z)^4$, don't go adding up 2 + 3 just because you see those exponents on the same base, y. That outside exponent, 4, must be dealt with first. The y^3z isn't allowed to come out and play until it deals with the 4. I mean, Ms. Exponent is telling y^3z to do something, and it had better obey before it interacts with anyone else. Those parentheses are supposed to be providing that sort of protection.

Doing the Math

Use the Power rule to simplify these. I'll do the first one for you.

1. $(-3d^5)^4$

<u>Working out the solution</u>: There's no addition or subtraction in this expression, so we can freely distribute the 4 exponent inside the parentheses. But we'd better watch the −3; if we just blindly distribute, we might end up with -3^4d^{20}, and that would be wrong. Why? Because the exponent, 4, is touching a set of parentheses that *includes the negative sign*, so the negative sign must also obey Ms. Exponent's command: "Multiply times yourself 4 times." So, we get $(-3d^5)^4 = (-3)^4d^{20} = 81d^{20}$.

Answer: $81d^{20}$

2. $(2ab^2)^4$

3. $-(2ab^2)^4$

4. $(-2ab^2)^4$

5. $\left(\dfrac{x^3y}{-5}\right)^3$

6. $(xy^2)(y^3z)^4$ *(Hint: See the Watch Out on p. 249.)*

(Answers on p. 409)

The Quotient Rule

With all the demands of her powerful job, Ms. Exponent often needs to take a break and do something relaxing like spend time with her adorable little kitten. Her kitty loves to run around the office, wrestle crumpled-up paper, and in her quieter moments, stare at the goldfish. As you might expect, though, this is no ordinary kitten. In fact, she even assists Ms. Exponent with a little copycat work of her own.

In Chapter 4, we canceled factors on fractions by finding hidden copycats—fractions with the same thing on top and bottom, which equal 1:

$$\frac{7a}{8a} \;=\; \frac{7}{8}\cdot\frac{a}{a} \;=\; \frac{7}{8}\cdot 1 \;=\; \boldsymbol{\frac{7}{8}}$$

But what if there's more than one power of a variable, like in $\dfrac{a^5}{a^2}$? We know $a^3 \cdot a^2 = a^5$, so we can rewrite the top of this fraction to pull out the biggest copycat we can, in this case, $\boldsymbol{\dfrac{a^2}{a^2}}$:

$$\frac{a^5}{a^2} \;=\; \frac{a^3\cdot a^2}{a^2} \;=\; \frac{a^3}{1}\cdot\boldsymbol{\frac{a^2}{a^2}} \;=\; a^3\cdot 1 \;=\; \boldsymbol{a^3}$$

Ta-da! We could also write this as "canceling" $\boldsymbol{a^2}$ from the top and bottom:

$$\frac{a^5}{a^2} \;=\; \frac{a^3\cdot \cancel{a^2}}{\cancel{a^2}} \;=\; \frac{a^3}{1} \;=\; a^3$$

Either way, we get $\dfrac{a^5}{a^2} = \boldsymbol{a^3}$. It's interesting that $5 - 2 = 3$. Hmm. Looks like we could have just *subtracted* the exponents. If instead we had started with $\dfrac{a^2}{a^5}$, do you see how we could pull out the same $\boldsymbol{\dfrac{a^2}{a^2}}$ copycat (or cancel $\boldsymbol{a^2}$ from the top and bottom) and end up with $\boldsymbol{\dfrac{1}{a^3}}$? Looks like we could have *subtracted* again. Time for a shortcut alert!

The Quotient* Rule for Exponents: AKA, the "Subtract the Exponents" Shortcut!

When you have variables in a fraction and those variables have *exponents*, look for *bases that are the same* and then cancel factors by subtracting exponents. So, if the exponents *m* and *n* are positive integers and *b* is a nonzero real number:

If $m \geq n$, then:

$$\frac{b^m}{b^n} = b^{m-n}$$

and if $m < n$, then:

$$\frac{b^m}{b^n} = \frac{1}{b^{n-m}}$$

This rule can look a little confusing, but just remember: When we have the same base on the top and bottom of a fraction, we look for whichever has the bigger exponent—and that's where the base will be left behind after we do the exponent subtraction.[†]

It's not so bad; after all, the whole point of this is to quickly find the copycats: They want to come out and play!

---

* I tell a story on p. 126 of *Math Doesn't Suck* to help you remember that *quotient* means "the answer in a division problem." (Remember, fractions *are* division!) The story has to do with Sparky the dog. Hmm, lots of pets around these parts.

† In the next chapter, when we learn about negative exponents, we'll see that *m* and *n* can actually be any values we want; we don't really need the $m \geq n$ and $m < n$ restrictions. Of course, *b* must <u>always</u> be nonzero.

Step By Step

Using the Quotient rule, AKA the "subtract the exponents" shortcut to reduce fractions:

Step 1. First, make sure that the top and bottom of your fraction are each a product of factors; in other words, they are stuck together with multiplication, not addition or subtraction.

Step 2. Write in 1 for any invisible exponents. This will help when it's time to subtract.

Step 3. Pick a base that appears on both the top *and* the bottom of the fraction, and look to see which one has the bigger exponent: the top or the bottom one?

Step 4. For each base, *subtract* their exponents and leave the base with its new exponent on the side of the fraction that had the *bigger* exponent.

And... Action! Step By Step In Action

Simplify $\dfrac{g^5 h^2}{g h^8}$.

Steps 1 and 2. There's no addition or subtraction. Great! And we can write in 1 for g's exponent on the bottom: $\dfrac{g^5 h^2}{g^1 h^8}$.

Steps 3 and 4. Notice that, in this case, we have *two* variables that appear on both the top and the bottom; that means we have *two* separate subtraction problems. For the base g, the bigger exponent is on top, so we get $g^{5-1} = g^4$ on the top. For h, the bigger exponent is on the bottom, so we get $h^{8-2} = h^6$ on the *bottom*. So, our fraction ends up looking like this: $\dfrac{g^4}{h^6}$. There's nothing else we can do, because we now have two different bases. (So don't go thinking we can subtract 6 – 4 or anything.)

Answer: $\dfrac{g^4}{h^6}$

Doing the Math

Simplify. (Assume no denominators equal zero.) I'll do the first one for you.

1. $\dfrac{3pq^3 \cdot p}{p^7}$

<u>Working out the solution</u>: The only base on the bottom is p, so p is the only base we'll be subtracting exponents with. It's got a power of 7 on the bottom. But what's its power on top? Hmm. Let's simplify this a little, first writing in our invisible exponents, and multiplying those two p's in the numerator together

by adding their exponents, $1 + 1 = 2$: $\dfrac{3p^1q^3 \cdot p^1}{p^7} = \dfrac{3p^2q^3}{p^7}$.

Now it's easier to see that we'll be subtracting p^{7-2}, ending up with p^5 on the bottom.

Answer: $\dfrac{3q^3}{p^5}$

2. $\dfrac{a^2b^4}{a^4b^2}$ 3. $\dfrac{w^5x^6}{8x^5w^7}$ 4. $\dfrac{3xy}{x}$

(Answers on p. 409)

Takeaway Tips

When you've got a bunch of stuff all multiplied and mushed together and exponents are involved, write it in "easy form." Go ahead, spread it out and write in the invisible 1 exponents.

Product rule: When two numbers or variables with the same base are being multiplied together, you can just *add* their exponents.

Power rule: When one exponent is being raised to another exponent, just *multiply* the two exponents together. And if parentheses are involved, only distribute the exponent if there is multiplication/ division inside, *not* addition/subtraction.

 Quotient rule: When two numbers or variables with the same base are being divided (in fraction form), just *subtract* their exponents. All we're doing is canceling copycats!

Gals Who Blog!

Think you're "just" a teenager and nobody cares about your opinion? Think again! Read these stories of gals just like you who decided to share their writing with the world.

BLOGGED SINCE THE AGE OF 8!

I started writing a blog when I was eight years old; I sort of followed suit when my mother started writing one. I would type articles on the computer and send them out to my parents' friends every week. I'm fourteen now, and I haven't missed a week. I write opinion pieces about current events, but I also write poetry and always include random funny quotes. I want people to learn something and also to be entertained.

As a kid it makes me feel more confident, because adults actually begin to take my opinion seriously. They see I've done the research and I know what I'm talking about. It's a great feeling to know that I'm worth listening to, and I just know the skills I'm developing will serve me later in life, too.

Gussie Roc, 14

HAPPINESS—THROUGH WRITING!

In grades 7 and 8, I focused all my energy on popularity and the social scene. I had good friends, and that should have been what mattered. But I wasn't in the "popular" crowd, and obsessing over it was making me miserable. In the eighth grade, I started writing in a journal—a lot. It always seemed to make me feel better because it felt like I was telling someone my problems. By the ninth grade, I was writing on a very regular basis—journal entries, short stories, you name it. I even convinced a local paper to let me write a student column. I was writing constantly, every day. My friends and parents were stunned by my new positive outlook on life and my new attitude: I'd found something that really mattered to me, and I was able to let go of the popularity issue that had seemed so important before.

By changing my goals and understanding what was really important to me, I became a happy person. But don't think it came easily. I didn't believe I could get published at 14 years old. But that didn't stop me from trying. I emailed probably 10 newspapers. I received only one positive reply—from the very last newspaper I emailed! It just goes to show that with each failure, if you don't give up, you can still achieve what you want to get! That's just something to think about for everyone. If you're not happy, maybe you're trying to achieve the wrong things...or maybe you're searching for happiness in the wrong location. If you look harder somewhere else, maybe you'll discover that it's there—and you never knew it!

Bailey Thompson, 15

Chapter 18

Keep Your Day Job
Negative (and Zero) Exponents

$\mathcal{W}$e've all seen those reality shows where people compete to get their dream job . . . that is, if they don't get eliminated or "fired" right there on the spot. Yikes! I don't know, man. Those shows seem really cruel to me. Anyway, Ms. Exponent is no lightweight herself when it comes to dealing with her employees (you know, numbers and variables).

In this chapter, we'll see exponents take new forms; they'll be zero and even negative. Ms. Exponent's no-nonsense ways prove you'd better stay on your toes if you work for her. Check it out:

The Neutralizer Exponent: Ms. Zero

Yes, exponents can be zero, for example, 3^0. What do you think the fate of 3 is? I mean, what does it even mean to have an exponent of zero? Hmm. Let's look at this pattern of descending powers of 3:

$$3^4 = 81 \qquad 3^3 = 27 \qquad 3^2 = 9 \qquad 3^1 = 3 \qquad 3^0 = ?$$

Okay—time to guess what 3^0 equals. Just take a guess. Nobody can hear your thoughts, not even me. Got an idea? Great!

There are a couple of common guesses for this. Most people think 3^0 should equal either **0** or **1**—something sort of neutral-ish. But we actually have the power to figure this out! Just for the heck of it, let's consider the expression $\frac{b^5}{b^5}$. It's a copycat fraction; we know this equals **1**, right? (Unless $b = 0$; then the fraction is undefined.) And if we use the "subtracting the exponents" rule from p. 252 to simplify $\frac{b^5}{b^5}$, we get $b^{5-5} = \boldsymbol{b^0}$. Hmm, $\frac{b^5}{b^5}$ equals $\boldsymbol{b^0}$ and it also equals **1**?

Well, lookie there! We've just discovered that **$b^0 = 1$** (unless $b = 0$), so that means **$3^0 = 1$**, too. We figured it out! (Did you guess correctly?) Indeed, for *all* nonzero* b,

$$b^0 = 1$$

Yeah . . . you don't want to mess with Ms. Exponent when she's in a bad mood; she has the ability to *neutralize* her victims. For example, $\left(\frac{1}{7}\right)^0 = 1$, $(-9)^0 = 1$, and $1375^0 = 1$. Even $(x - y)^0 = 1$, no matter what x and y equal,[†] because Ms. Zero Exponent is touching the parentheses, which means she's touching everything inside them. Neutralized.

Ms. Exponent has also been known to demote people to lower floors. Generally speaking, in a big, tall office building like hers, the more prestigious the job is, the higher the floor, so you get the picture. And when she demotes someone? Let's just say the experience is rather *negative*.

Negative Exponents

We're going to get wild and wacky now:

$$2^{-1}$$

"Um, excuse me?" you might say. "We were just getting comfortable with exponents, and now you go and make them negative?" Yes, and you're going to be just fine. However, Mr. 2 might not be so fine. He's about to be demoted to a lower floor.

$$2^{-1} = \frac{1}{2}$$

Those two expressions are *equal* to each other, which is why an equals sign is between them. They represent the *same value*. In fact, for all nonzero b,

$$b^{-1} = \frac{1}{b}$$

Hmm . . . Do we really just have to "believe" this definition of the -1 exponent? Is there any logical reason why this would be true? (I'm glad you asked . . .)

. . . .

* We have to say "for nonzero b" because the expression 0^0 is *undefined* in math—just like it's *undefined* for a fraction's denominator to be zero. Think about it: Should 0^0 equal zero or 1? I mean, zero times *anything* is zero, but on the other hand, *everything else* to the zero power equals 1. Don't let this keep you up at night.
† Unless $x = y$, in which case we'd have 0^0; see previous footnote.

Negative Exponents: Why, oh Why?

Just keep an open mind, okay? Look at the pattern in the chart below: Left to right, as 2's exponents go from $3 \rightarrow 2 \rightarrow 1 \rightarrow 0$, the values get *cut in half* each time: $8 \rightarrow 4 \rightarrow 2 \rightarrow 1$. So then, as the exponents *continue* to decrease from $0 \rightarrow -1 \rightarrow -2 \rightarrow -3$, shouldn't the values *continue* getting cut in half? They'd go from $1 \rightarrow \frac{1}{2} \rightarrow \frac{1}{4} \rightarrow \frac{1}{8}$.

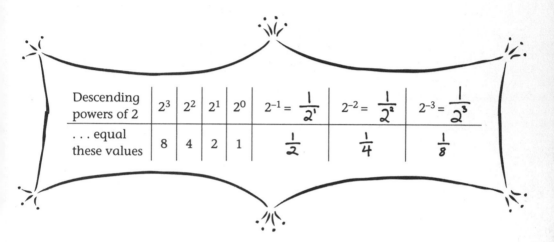

Descending powers of 2	2^3	2^2	2^1	2^0	$2^{-1} = \frac{1}{2^1}$	$2^{-2} = \frac{1}{2^2}$	$2^{-3} = \frac{1}{2^3}$
. . . equal these values	8	4	2	1	$\frac{1}{2}$	$\frac{1}{4}$	$\frac{1}{8}$

And yep, the pattern continues, and there is justice in the world. After reading this chart a few times, you see how negative powers start to have some meaning. Makes me feel better, anyway.

Notice in the last column that if Mr. 2 was supposed to be multiplying times himself 3 times, but he's getting demoted like this, 2^{-3}, then he *still* has to do that job—he just has to do it from a lower floor:

$$2^{-3} = \frac{1}{2^3}$$

In fact, for all nonzero b and all real numbers m, the following is true:

$$b^{-m} = \frac{1}{b^m}$$

No matter what b is (as long as it isn't zero), a negative exponent will send it downstairs. For example, $286^{-5} = \frac{1}{286^5}$ and $(0.3)^{-2} = \frac{1}{(0.3)^2}$.

But there's a positive "flip" side to all this negative exponent business. Let's say Mr. 2 is downstairs doing a spectacular job. Then Ms. Exponent might be outraged, saying, "Hey, you're better than this piddly position. I'm promoting you: Get your butt <u>upstairs</u>!" And for grateful Mr. 2, this could mean things like:

$$\frac{1}{2^{-1}} = 2 \quad \text{and} \quad \frac{1}{2^{-3}} = 2^3$$

In fact, for all nonzero b and all real numbers m:

$$\frac{1}{b^{-m}} = b^m$$

QUICK NOTE   We can always write b^m and b^{-m} over 1, which looks more like "upstairs." Then we could write:
$$\frac{b^{-m}}{1} = \frac{1}{b^m} \quad \text{and} \quad \frac{1}{b^{-m}} = \frac{b^m}{1}.$$

Even when a bunch of variables and numbers are mixed together in a fraction, we can still use the upstairs/downstairs trick. Check it out:

Shortcut Alert

For fractions whose top and bottom are each a product of factors (so no addition or subtraction is involved), any time we see a negative exponent on a base, we can just say, "Hey, you're on the wrong floor!" Then *flip its position in the fraction* and get rid of the exponent's negative sign. So, for example:

$$\frac{6^{-1}}{7^{-1}} = \frac{7}{6} \qquad \frac{11}{y^{-2}} = 11y^2 \qquad \frac{8^{-1}a}{b^{-3}} = \frac{ab^3}{8} \qquad \frac{(-n)^{-5}}{k^{-1}} = \frac{k}{(-n)^5}$$

Notice that only the *exponent's* negative sign—and no other negative signs—go away when we change floors.

Watch Out!*

If addition or subtraction is involved, be careful when you're flipping. For example, this *is* true: $\dfrac{1}{(x-y)^{-1}} = x - y$. You could have substituted the entire $(x - y)$ for a variable like b, and then you'd see how it's a simple application of our rules. However, $\dfrac{1}{(x^{-1} - y^{-1})}$ does NOT equal $x - y$. In fact, those parentheses aren't even doing anything, so we can drop them. And since $x^{-1} = \frac{1}{x}$ and $y^{-1} = \frac{1}{y}$, we can actually rewrite this as $\dfrac{1}{x^{-1} - y^{-1}} = \dfrac{1}{\frac{1}{x} - \frac{1}{y}}$, which we simplified on p. 66 and got $\dfrac{xy}{y-x}$.

Doing the Math

Simplify as much as you can, and don't leave any zero or negative exponents behind. (Assume all variables are nonzero.) I'll do the first one for you.

1. $c^{-1} + 0.5b^0 - (3a)^{-2}$

<u>Working out the solution</u>: We'll take these terms one at a time. First, we know that $c^{-1} = \frac{1}{c}$, right? Next, since b has a zero exponent, $b^0 = 1$. (Notice that the 0 exponent is not touching the 0.5, though!) Finally, we'll flip the third term, so our expression looks like $\frac{1}{c} + 0.5 - \frac{1}{(3a)^2}$, which is great because we've gotten rid of all the zero and negative exponents. Now we just finish by squaring the 3a, and with no like terms to combine, we're done!

Answer: $\dfrac{1}{c} + 0.5 - \dfrac{1}{9a^2}$

. . . .
* Assume in this paragraph that $x \neq 0$, $y \neq 0$, and $x \neq y$. Just so everything is legal!

2. $\dfrac{1}{2^{-4}}$

3. $\dfrac{(e^0 + f^0)f^{-1}}{e^{-3}}$

4. $x[3x^0 - (3x)^0]^2$ (Hint: We can't distribute that 2, so simplify the inside first.)

5. $(6xz^{-2})^0 + (5x^0z)^{-2}$

6. $\dfrac{2^{-3}}{a^{-2}}$

7. $\dfrac{1}{(c^{-3} + d^0)^{-1}}$ (Hint: Imagine $(c^{-3} + d^0)$ replaced by w. How would you start?)

(Answers on p. 409)

What's The Deal

$\mathcal{I}$'m going to let you in on a little secret: The Quotient rule (p. 252) is actually the same rule as the Product rule (p. 241) but with negative exponents! Let's multiply $n^5 \cdot n^{-2}$ using the Product rule, so we'll just add the exponents:

$$n^5 \cdot n^{-2} = n^{5 + (-2)} = \mathbf{n^3}$$

To use the Quotient rule on this same problem, we'd first get rid of the negative exponents by writing it as a fraction and then *subtract* exponents:

$$n^5 \cdot n^{-2} = \frac{n^5}{n^2} = n^{5 - 2} = \mathbf{n^3}$$

That's right, the "subtracting the exponents" shortcut is just the "*adding the exponents*" shortcut in disguise. And it makes sense when you think about it, since subtracting is the same thing as adding negative numbers.* All this math stuff is connected!

. . . .

* For a review of this, see Chapter 1 in *Kiss My Math*.

The Product and Power Rules . . .
with Negative Exponents

Using the Product rule, when we multiply two similar bases that *both* have negative exponents, it's not surprising that the answer's exponent will still be negative, because we're just adding two negative exponents together, right?

$$(2^{-2})(2^{-4}) = 2^{-2+(-4)} = 2^{-6} = \frac{1}{64}$$

On the other hand, with the Power rule (see p. 246), when we raise a negative exponent to another negative exponent, since we *multiply* the negative exponents together, we end up with a <u>positive</u> exponent. For example:

$$(2^{-2})^{-4} = 2^{-2 \cdot -4} = 2^{8} = 256$$

Crazy! Let's simplify this *without* the Power rule and see what's going on. We'll start by dealing with just the inside of the parentheses. Read each step *slowly*; you can do this:

$$(2^{-2})^{-4} = \left(\frac{1}{2^{2}}\right)^{-4} = \left(\frac{1}{4}\right)^{-4} = \frac{1^{-4}}{4^{-4}} = \frac{4^{4}}{1^{4}} = 4^{4} = 256$$

Same answer. Of course, this was easier to simplify using the Power rule. Yep, Ms. Exponent's got it all figured out.

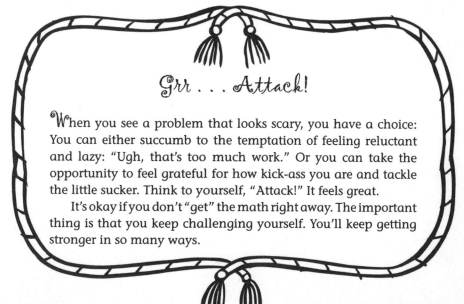

Grr . . . Attack!

When you see a problem that looks scary, you have a choice: You can either succumb to the temptation of feeling reluctant and lazy: "Ugh, that's too much work." Or you can take the opportunity to feel grateful for how kick-ass you are and tackle the little sucker. Think to yourself, "Attack!" It feels great.

It's okay if you don't "get" the math right away. The important thing is that you keep challenging yourself. You'll keep getting stronger in so many ways.

Invisible Exponents and Reciprocals

We've been using invisible 1 exponents, but notice that *we could leave them on* if we wanted, which is helpful when we use the Product and Quotient rules. So:

$$b^{-1} = \frac{1}{b^1} \qquad \text{and} \qquad \frac{1}{b^{-1}} = b^1$$

And by the way, applying the −1 exponent *twice* will bring you back to where you started! For example, $(5^{-1})^{-1} = 5^{-1 \times -1} = 5^1 = 5$. I don't know about you, but this reminds me a lot of reciprocals . . . and for a good reason:

QUICK NOTE Applying an exponent of −1 is like saying "the reciprocal of."* So, for example, 5^{-1} means "the reciprocal of 5," so $5^{-1} = \frac{1}{5}$. Similarly, $\left(\frac{1}{5}\right)^{-1}$ means "the reciprocal of $\frac{1}{5}$," so $\left(\frac{1}{5}\right)^{-1} = 5$. One more: $\left(\frac{3}{2}\right)^{-1} = \frac{2}{3}$. Notice that we could just distribute the exponent, and get the same answer: $\left(\frac{3}{2}\right)^{-1} = \frac{3^{-1}}{2^{-1}} = \frac{2}{3}$.

Mixing It All Together

With so much going on, what's the best way to attack something like this?[†]

$$16x^{-1}(16^{-1}x + 2^{-1}x^2)$$

I recommend first rewriting everything so it has positive exponents and going from there. <u>Somebody on the wrong floor? Let's take care of 'em!</u>

$$16x^{-1}(16^{-1}x + 2^{-1}x^2) = \frac{16}{x}\left(\frac{x}{16} + \frac{x^2}{2}\right)$$

. . . .

* To review reciprocals (re-FLIP-rocals!), check out Chapter 5 in *Math Doesn't Suck*.
[†] Assume $x \neq 0$.

Now we distribute: $\frac{16x}{16x} + \frac{16x^2}{2x}$. Reducing, we get **1 + 8x**. Ah, much simpler.

Answer: 1 + 8x

Step By Step

Simplifying complicated exponential expressions:

Step 1. First, I recommend dealing with all the negative exponents. Make 'em positive and get 'em (and their bases) on the correct side of the fraction line!

Step 2. Simplify what you can. If everyone's connected only by multiplication and division, you can use the **Product and Quotient rules** to add/subtract exponents on similar bases (writing the invisible exponents as 1 will help), and the **Power rule** to distribute exponents over factors inside parentheses.

Step 3. Keep using the exponent rules to cancel factors and simplify until there aren't any negative or zero exponents left and everyone's reduced. Done!

And... Action! Step By Step In Action

Simplify $\left[\dfrac{2^{-1}a^5}{2a^{-1}b^{-1}} \right]^{-3}$ *(assume a, b $\neq$ 0).*

Steps 1 and 2. Let's start with what's *inside* the big brackets first and pretend the −3 exponent isn't even there for now. By the way, don't go canceling anything yet (those 2's won't actually cancel!). Rewriting just the *inside* to get rid of its negative exponents by flipping the positions of their bases, we get $\dfrac{2^{-1}a^5}{2a^{-1}b^{-1}} = \dfrac{a^5 \cdot a^1 \cdot b^1}{2 \cdot 2^1} = \dfrac{a^6 b^1}{4}$. The entire problem now looks like this:

$$\left[\dfrac{a^6 b}{4} \right]^{-3} .$$

Step 3. Now, let's use the Power rule to distribute the –3 inside: $\left[\dfrac{a^6b}{4}\right]^{-3} = \dfrac{a^{-18}b^{-3}}{4^{-3}}$. Seems everyone's on the wrong floor! So we get

$$\dfrac{a^{-18}b^{-3}}{4^{-3}} = \dfrac{4^3}{a^{18}b^3} = \dfrac{64}{a^{18}b^3}.$$

Answer: $\dfrac{64}{a^{18}b^3}$

QUICK NOTE   When a big messy fraction is raised to a negative power, you can deal with the inside first, like we did above, or start by flipping everything, exactly as it is, like this:

$$\left(\dfrac{\text{big mess}}{\text{another big mess}}\right)^{-5} = \left(\dfrac{\text{another big mess}}{\text{big mess}}\right)^{5}$$

It's your choice! You'll get the same answer either way.

"*When I get really confused in my math homework, I just have to stop what I'm doing, take a breath, and remember it's not the end of the world. Sometimes I take like a 10-minute break to go for a quick run, and when I come back and try again, things usually make more sense.*" **Isabella, 14**

So Many Negatives—My Head's Spinning!

What happens when we have a negative number raised to a negative power, like $(-4)^{-3}$? Don't panic. We'll take it one step at a time: First, we just put the entire $(-4)^3$ downstairs, and then it's not so scary. We can finish as we normally would.

$$(-4)^{-3} = \dfrac{1}{(-4)^3} = \dfrac{1}{-64} = -\dfrac{1}{64}$$

Remember, the negative sign on the exponent only means that we flip that term to the other side of the fraction; <u>a negative exponent has nothing to do with the *sign* of the answer</u>. By the way, if we have an *even* power, then the negative sign on the 4 will indeed go away, like this: $(-4)^{-2} = \dfrac{1}{(-4)^2} = \dfrac{1}{16}$. However, if the exponent's *inside* like this, (-4^{-2}), notice that the exponent is NOT touching the negative sign on the 4. Those first parentheses aren't even doing anything now!

$$(-4^{-2}) \;=\; -4^{-2} \;=\; (-1)\cdot 4^{-2} \;=\; (-1)\!\left(\dfrac{1}{4^2}\right) \;=\; -\dfrac{1}{16}$$

Every case is different. Just pay attention to what's touching what, remember what negative exponents *mean*, and you'll be fine. Read this section a couple of times; it's a lot to take in. I'm very proud of you for hanging with it. Seriously, you rock.

QUICK NOTE <u>I recommend getting rid of negative exponents as your very first step</u>, but I'll show you a different strategy in problem 1 below, where nothing looks like a fraction until the very end.

Doing the Math

Simplify the following expressions, and don't leave any negative or zero exponents (assume no denominators equal zero). I'll do the first one for you.

1. $[q(-3p^{-7}p^5)^4]^{-1}$

<u>Working out the solution:</u> Let's ignore the outside -1 exponent for now and just work inside the brackets: $q(-3p^{-7}p^5)^4$. Notice that we cannot move that q inside the parentheses, because we don't want q to end up with the 4 exponent on it! Let's first

multiply the p terms by adding their exponents, and since $-7 + 5 = -2$, we get $q(-3p^{-7}p^5)^4 = q(-3p^{-2})^4$. Now let's distribute the 4 [notice that $(-3)^4 = 81$], and we get $q(-3p^{-2})^4 = q(81p^{-8}) = 81p^{-8}q$. The whole problem has become: $[81p^{-8}q]^{-1}$. Distributing the -1 exponent, we get $81^{-1}p^8q^{-1}$. Finally, we'll rewrite it without negative exponents:

$81^{-1}p^8q^{-1} = \dfrac{p^8}{81q}$. Phew!*

Answer: $\dfrac{p^8}{81q}$

2. $[(6^1)(7^{-1})]^{-1}$

3. $(32b^{-6})(2b^{-5})$

4. $\left(\dfrac{(-4)^{-2}}{2^{-4}}\right)^{-1}$

5. $\left[\dfrac{w^3(2 + x)}{w^7(1 - y)}\right]^{-1}$ *(Hint: See the QUICK NOTE on p. 266.)*

6. $\left(\dfrac{3q^{-1}}{3^{-1}}\right)^{-2} \div \dfrac{1}{q}$ *(Hint: See the middle of p. 43.)*

(Answers on p. 409)

Scientific Notation

 O ne application for negative exponents is called *scientific notation,* which uses positive and negative exponents to express really big and really small numbers. Hmm. Perhaps the big company Ms. Exponent runs is an engineering firm.

* * * * * * * * * *

* Perhaps now you see why I prefer to get rid of negative exponents first.

What's It Called?

Scientific notation is a helpful way to express really big and really small numbers, where we put the decimal point after the first nonzero digit of the number and then multiply it times the necessary power of 10.

Here are some examples of scientific notation:

2.0 $\times$ **10^{-5}** (*thickness, in meters, of a piece of glitter*)

. . . looks better than 0.00002

7.5 $\times$ **10^{18}** (*approximate number of grains of sand on Earth*)*

. . . looks way better than 7,500,000,000,000,000,000!

To write a number in scientific notation, we just take the decimal point and move it enough places until the number is between 1 and 10 (but not equal to 10), and then the <u>number of places we moved it</u> becomes the positive or negative exponent on 10. Draw little loops as you count to keep track of how many places you've moved. For example:

$$0.00002 = 2.0 \times 10^{-5}$$

moved to the right, 5 places

$$7,500,000,000,000,000,000 = 7.5 \times 10^{18}$$

moved to the left, 18 places

To add, subtract, multiply, or divide these crazy-looking numbers, I like to rewrite the 10 as a variable like *b*, and then just use the exponent rules as normal.

. . . .

* According to www.hawaii.edu.

For example, how could we evaluate this?

$$(3.2 \times 10^{-6})(2.0 \times 10^{-5})$$

If this looks confusing, we can just stick in b for 10 and rewrite it as $(3.2b^{-6})(2.0b^{-5})$. Now we can multiply the b's together by adding exponents, and we get:

$$(3.2)(2)b^{(-6)+(-5)} \ = \ 6.4b^{-11} \ = \ 6.4 \times b^{-11}$$

And since $b = 10$, we know our real answer is **6.4×10^{-11}**.

This works for other operations, too. I think sticking in variables makes scientific notation much easier to swallow. I mean, who says we can't be a little sneaky and use variables to make *numbers* easier from time to time?

For more details and some practice problems on scientific notation, check out danicamckellar.com/hotx.

 Takeaway Tips

 Zero exponents neutralize whatever base they're touching:* Their base becomes 1.

Negative exponents don't affect the *sign* of the answer.

An exponent of –1 means "the reciprocal of."

Negative exponents mean we're on the wrong side of the fraction line. As long as everything's stuck together with multiplication, just take the negative sign off the exponent and move it and its base to the other side.

Exponents affect only what they are *touching*. If they are touching parentheses, then they affect everything inside the parentheses.

. . . .

* . . . unless the base is 0, since 0^0 is undefined.

TESTIMONIAL

Morgan Clendenen
(Burlington, NC)
<u>Before</u>: Embarrassed and frustrated student
<u>Today</u>: Wine maker extraordinaire!

When I was younger, math was always a struggle for me! It was especially embarrassing because my mother was a calculus teacher at my high school. I felt like a math idiot no matter how hard I tried. I also had dyslexia, so I was constantly transposing figures and numbers. In algebra, I would do all this work on equations but get the wrong answers over and over again.

Because I struggled in math, I clung to the idea that "I'm never going to use this" so I wouldn't have to feel bad. When I finished the last math course I ever took, I remember being overjoyed. I thought, "I'm never going to have to do math again!" Then I got my first job selling brick and found myself doing geometry problems to figure out how much brick each job needed! For me, in the end, the key was practice—and I got that practice in the real world.

"If you have a dream, go for it."

Today, I'm a wine maker. I started out as a wine sales representative but always had this vision of myself owning my own vineyards—and now I do! I own a company that produces wine from grapes and then ships it all over the globe. A lot of the products that I use are European, so they're based on the metric system. I'm constantly converting between currencies (euros vs. dollars), degrees (Celsius vs. Fahrenheit), and measurement figures (gallons vs. liters). If you want to work in an international business of any kind (or any business, for that matter), math is so important! I have a French business partner, so I have to do these types of conversions all the time.

I also work with exponents, because acidity levels are measured in terms of pH, which is very important when picking grapes. You've probably heard of pH, but did you know that the abbreviation may have come from a translation of "*power* of **H**ydrogen"? This is because the pH is actually an *exponent*; in fact, it's an exponent of 10. You probably know that $10^8 = 10 \times 10^7$, so 10^8 is ten times bigger than 10^7, right? Well, pH works just the same way. But the *lower* the pH, the *higher* the acid level, so that means that a wine with a pH level of 3 would be 10 times more acidic than a wine with a pH level of 4. If your wine is outside the bracket of (3.4, 3.7), you're in trouble though!

Before I got involved in the business, I thought, "Okay, you pick some grapes, mash them up, put them in bottles, and sell the bottles to stores." In reality, wine making is much more involved and incredibly gratifying. I'm making a product that stands the test of time, and I get to eat a lot of amazing food and drink a lot of good wine on the job!

The truth is, you don't have to be the best in your math class. You really just need to do your best to build a strong foundation. While this might take some hard work, most things in life worth having are worth working for! So don't get bogged down in the "I can't," but rather revel in small victories. If you have a dream, go for it. I grew up in North Carolina, far from any vineyards, but I had this image of myself out in the fields of Napa, picking grapes, making wine—and I did it!

Chapter 19

The Benefits of a Diary

Intro to Square Roots and Radical Expressions

Do you write in a journal? I do. It's always helped me, especially when my emotions run wild. No matter how irrational my thoughts seem when they're trapped in my head, writing them down helps me think more clearly, and I feel so much better.

Right now, it's time to *decorate* your journal. Let's say there's a tilted square on the front, and so far, you've lined one side with 3 big, sparkly sticker jewels. How many total jewels will fill the square? If you fill them in, you'll see that the answer is 9:

$$3^2 = 9$$

This reads, "Three squared equals nine." In other words, 9 is the **square** of 3.

Now, let's go backward. If we know the area of the square is covered by 9 jewels, then how many would go on just one side? The answer is 3, and this is because:

$$\sqrt{9} = 3$$

This reads, "The square root of 9 equals 3." Because 9 jewels can be arranged into a square (with regular rows and columns), it's called a **perfect square**, or a square number. And the length of one side of that square is called the **square root**. Yep, all these terms have the word *square* in them. Let's sort this out!

What's It Called?*

The **square** of a number is that number, times itself. For example:

The *square* of 4 is 16 because $4 \cdot 4$ or $4^2 = 16$.

The *square* of $\frac{1}{2}$ is $\frac{1}{4}$ because $\frac{1}{2} \cdot \frac{1}{2}$ or $\left(\frac{1}{2}\right)^2 = \frac{1}{4}$.

The *square* of 0.1 is 0.01 because $(0.1)(0.1)$ or $(0.1)^2 = 0.01$.

The *square* of (-4) is 16 because $(-4)(-4)$ or $(-4)^2 = 16$.

For all real numbers b, the square of b is b^2. Squaring a number is like saying "What's the *area* of a square whose sides are this length?"

Notice that when we *square* a number, we get its square. We just used the word *square* first as a verb and then as a noun. Good times!

Of course, instead of using sparkly jewels, we could also measure using inches. So if a square's sides have a length of 3 inches each, its area must be 9 square inches. (Yep, we just used *square* as an adjective.)

By the way, it has always seemed crazy to me that you can have a square shape whose sides have a length of $\frac{1}{2}$ inch and yet its area is $\frac{1}{4}$ square inch.[†] It seems like the area is somehow smaller than one of its sides, right? But that's like comparing apples to oranges, because the *units* are different: The area is expressed in *square*

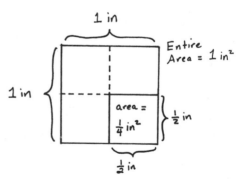

inches (in.²), which I guess is a much bigger deal than just plain 'ol inches (in.). Look at the divided 1 inch square above and see how it actually sort of makes sense why a square of length $\frac{1}{2}$ in. should have an area of $\frac{1}{4}$ in.².

Now, how about these *square roots*?

. . . .

* To review multiplication of decimals, see Chapter 10 in *Math Doesn't Suck*.
† By the way, *any* time we square a positive number that is less than 1, we'll get a smaller number than we started with. Try a few and check it out.

What's It Called?

Ring Ring...

The **square root** of a number is the *positive* value (or zero), which when multiplied times itself gives you the original number. So, finding the square root is the inverse action of squaring a positive number (or zero). For example:

$$\sqrt{4} = 2 \quad \text{because} \quad 2^2 = 4$$

$$\sqrt{49} = 7 \quad \text{because} \quad 7^2 = 49$$

$$\sqrt{0.0004} = 0.02 \quad \text{because} \quad (0.02)^2 = 0.0004$$

$$\sqrt{\frac{1}{49}} = \frac{1}{7} \quad \text{because} \quad \left(\frac{1}{7}\right)^2 = \frac{1}{49}$$

$$\sqrt{0} = 0 \quad \text{because} \quad 0^2 = 0$$

And, in fact, for all $a \geq 0$:*

$$\sqrt{a^2} = a$$

And this makes sense. We're just saying that $\sqrt{7^2} = 7$.

The square root sign $\sqrt{}$ sort of looks like it could be enclosing the shape of a square inside it, right? And it's saying "Get this big square out of me!" So you huff and you puff and with all your might, you pull out the square. But it collapses on its way out, and all you're left with is the length of one side of the (pretend) square. And that *is* the value of the square root. Try to pull out 25, and all you're left with is 5. Try to pull out 49, and you're left with 7.

Another way to think of it is this: Finding the square root is like saying, "See that number inside of me? That's the area of a square. What's the length of one of its sides?"

I like all the drama of trying to pull out the big square though.

. . . .

* This is similar to the formal definition of the square root of positive numbers, which you'll use later in math: For $a \geq 0$, $(\sqrt{a})^2 = a$. This just means that, by definition, $(\sqrt{5})^2 = 5$. No big deal.

Watch Out!

The square root of a negative number is *not defined* for the set of real numbers.

$$\sqrt{-16} = ?$$

I mean seriously, can you think of a real number that when multiplied times itself will give you −16? I didn't think so.*

What's the Deal?

Okay, listen up. Even though it's true that $(-4)^2 = 16$, **square roots of numbers are <u>defined</u> to always be positive** (or zero), so $\sqrt{16} = 4$. If you said that $\sqrt{16} = -4$, it would be wrong. Weird, huh? It's just a definition thing so that no square root expression can have more than one value.[†]

QUICK NOTE   "<u>Take</u> the square root of 25" and "<u>Find</u> the square root of 25" mean the same thing.

Cube Roots and Beyond

In addition to square roots, there are also cube roots, fourth roots, and beyond. A cube root of a positive number asks the question, "See that number inside of me? It's the volume of a cube; what's the length of one of its sides?"

. . . .

* But an *imaginary number* called $4i$ would do the job. More on that in Algebra II!
† Advanced footnote: In fact, this rule of square roots always being positive (or zero) makes $f(x) = \sqrt{x}$ a *function* and not just a relation! (See Chapter 9 for information about functions.)

Below, notice the tiny 3 in the corner. That means it's a cube root and not a square root. To evaluate a cube root, we just find the number that when raised to the third power gives the number inside. For example:

$$\sqrt[3]{8} = 2 \quad \text{because} \quad 2^3 = 8, \quad \sqrt[3]{27} = 3 \quad \text{because} \quad 3^3 = 27,$$

$$\sqrt[3]{64} = 4 \quad \text{because} \quad 4^3 = 64$$

For fourth roots, we put a little 4 up in the corner, and so on. For example, $3^4 = 81$, so we know that $\sqrt[4]{81} = 3$. And because $2^5 = 32$, we know that $\sqrt[5]{32} = 2$.

By the way, cube roots (and all "odd" roots) *can* involve negative numbers. If the inside is negative, then the answer is negative, and vice versa. For example, $\sqrt[5]{-32} = -2$ because $(-2)^5 = -32$, and $\sqrt[3]{-27} = -3$ because $(-3)^3 = -27$.

QUICK NOTE $\sqrt[2]{\ }$ is the same exact symbol as $\sqrt{\ }$. But people don't usually bother to write in the little 2, so if there's no little number, you'll know it's a square root!

The more general name for a square root symbol (and cube root symbol, etc.) is **radical sign**.

𝒲hat 𝒜re 𝒯hey 𝒞alled?

Radical sign is another name for the square root sign, $\sqrt{\ }$. (It can also mean $\sqrt[3]{\ }$, $\sqrt[4]{\ }$, etc.) To some people, *radical* means "like, totally cool." I mean, I guess these symbols are kinda cool, if you're into that sorta thing.

A **radical expression** (or **radical** for short) is a math expression that contains at least one radical sign. These are both examples of *radical expressions*:

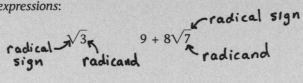

The **radicand** is whatever is underneath the radical sign. And how can we remember the word *radicand*? Well, it's not enough to have the radical sign, after all. We need the radical sign **and** something underneath it: the radic**and**. Get it?

Watch Out!

Don't confuse the word *radical* with the word *rational*. <u>Rational</u> has to do with values that can be expressed as <u>ratios</u> of whole numbers, like fractions.

Some Good Roots to Know

Common square roots: $\sqrt{4} = 2$; $\sqrt{9} = 3$; $\sqrt{16} = 4$; $\sqrt{25} = 5$; $\sqrt{36} = 6$; $\sqrt{49} = 7$; $\sqrt{64} = 8$; $\sqrt{81} = 9$; $\sqrt{100} = 10$; $\sqrt{121} = 11$; $\sqrt{144} = 12$

Common cube roots: $\sqrt[3]{8} = 2$; $\sqrt[3]{27} = 3$; $\sqrt[3]{64} = 4$; $\sqrt[3]{125} = 5$

Common fourth roots: $\sqrt[4]{16} = 2$; $\sqrt[4]{81} = 3$

Common fifth root: $\sqrt[5]{32} = 2$

✏️ Doing the Math

Write the following expressions without using radical signs. If the expression does not represent a real number, write "not a real number." I'll do the first one for you.

1. $\sqrt[3]{-27} - \sqrt{0.09} + \sqrt{-27}$

Working out the solution: Let's take this one term at a time. Because $(-3)^3 = -27$, we know that $\sqrt[3]{-27} = -3$. The next term is a square root, and since $(0.3)^2 = 0.09$, we know that $\sqrt{0.09} = 0.3$. So far, our expression looks like this: $-3 - 0.3 + \sqrt{-27} \rightarrow -3.3 + \sqrt{-27}$. But wait, that last term has a negative sign under the square root, and that does not represent a real number. So the entire expression, as a whole, does not represent a real number.

Answer: Not a real number

2. $\sqrt[3]{-1} + \sqrt[3]{1} + \sqrt[3]{125}$

3. $\sqrt{81} - \sqrt[4]{81} + \sqrt{0.0081}$

4. $\sqrt{-1} - \sqrt[3]{1}$

5. $\sqrt[3]{-64} - \sqrt[3]{-27} + \sqrt[3]{0.001}$

(Answers on p. 409)

From this point on, we'll deal with square roots only.

Pythagorean Theorem

Square roots can take us beyond square shapes to the land of triangles and rectangles, too. See, a long time ago in an ancient land, someone figured out the relationship between the 3 sides of a right triangle. For a right triangle* with short sides of lengths a and b and longest side (hypotenuse) of length c, this is always true:

$$a^2 + b^2 = c^2$$

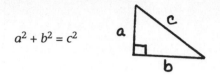

This means if we know the length of two sides of a right triangle, we can figure out the length of the third side. For example, if the shorter sides are 3 inches and 4 inches, then:

$$3^2 + 4^2 = c^2$$

$$\rightarrow 9 + 16 = c^2$$

$$\rightarrow c^2 = 25$$

(taking the square roots of both sides)

$$\rightarrow \sqrt{c^2} = \sqrt{25}$$

$$\rightarrow c = \mathbf{5}^\dagger$$

And we just learned the hypotenuse's length!

Reality Math

Pool Party Pizzazz

You're in charge of decorating for a pool party, and you want to string lights diagonally across the pool, hanging a beautiful lantern in the center.

You've got poles at two corners, but how long should the string of lights be? You know the pool is 8 yards long and 6 yards

---

* A *right triangle* is a triangle that has one right angle (90 degrees).
† Technically, $c = 5$ or -5, but for the physical world, only 5 makes sense.

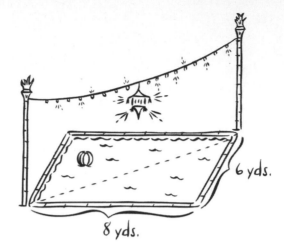

wide, but for the moment, you're by yourself with no quick way to measure the diagonal of the pool.

No problem for a math-savvy gal like yourself! On a party napkin, you quickly sketch the diagram and easily find the diagonal length you need:

$$6^2 + 8^2 = d^2$$

$$\rightarrow \quad 36 + 64 = d^2$$

$$\rightarrow \quad 100 = d^2$$

(taking the square roots of both sides)

$$\rightarrow \sqrt{100} = \sqrt{d^2}$$

$$\rightarrow d = \mathbf{10^*}$$

You need 10 yards of lights (plus a few extra inches so the light hangs nicely). Problem solved. And you know what? The pool looks gorgeous.[†]

---

* Technically, $d = 10$ or -10. But for the physical world, only 10 makes sense. (We'll do more with square roots of *variables* in Chapter 25.)
† Don't ever suspend electricity over water without adult supervision. You could electrocute yourself! (That's usually fatal, by the way.)

Irrational Numbers

Yes, things are about to get crazy. You see, there won't always be perfect squares under the square root signs, like for example, $\sqrt{2}$ and $-3\sqrt{5}$.

So what *is* $\sqrt{2}$? Hmm. This funny looking thing does have a value. It can be graphed on the number line and everything. So, what *value*, times itself, gives us 2? Turns out that it's approximately 1.41421, but the exact value of $\sqrt{2}$ simply can't be written as a fraction or as a terminating or repeating decimal. That's right; it's an *irrational number*, just like π. Check out the chart on p. 401 for approximate values of some irrational square roots.

How about the value of $-3\sqrt{5}$? According to the above referenced chart, $\sqrt{5} \approx 2.236$, so that means $-3\sqrt{5} \approx -3(2.236) \approx -6.708$.

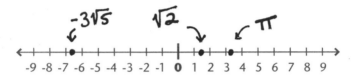

Just because we can't write out all the decimal places of irrational numbers doesn't mean they don't have a value, after all! Irrational numbers are people, too. The truth is, when dealing with irrational numbers in the form of square roots, unless you've been asked for an approximation, I find it best to leave them in square root form—as long as they are *simplified*, which we'll learn to do in a moment.

Betcha didn't know that writing in a journal can help with our emotions *and* with radicals. . . .

best advice, even when they're trying. For me, writing in my private journal was the best way to go. If I could just get my feelings out on the page, I could usually separate the calm, level-headed, rational thoughts from the immature, irrational ones. Suddenly I could think more clearly, I understood my feelings better, and everything just seemed a lot simpler. I highly recommend it. (If you're worried about someone else reading what you write, check out pp. 122-23 for an idea.)

The same strategy works for simplifying radical expressions: We'll begin by separating the rational stuff from the irrational stuff, and then things will get, well, simpler!

QUICK NOTE   When you see a number pressed up against a radical, it means they're being multiplied together, just like with variables:

$$4\sqrt{5} = 4 \cdot \sqrt{5} \text{ and } 4y = 4 \cdot y$$

Simplifying Radical Expressions:
Looking for Perfect Squares

If we saw something like $\sqrt{18}$, we might say, "Oh, well, 18 isn't a perfect square. Better luck next time." But wait! There might be some rational stuff we can salvage. Because 9 is a factor of 18 and it is also a perfect square, look at how we can rewrite this:

$$\sqrt{18} = \sqrt{9 \cdot 2} = \sqrt{9} \cdot \sqrt{2} = 3 \cdot \sqrt{2} = \mathbf{3\sqrt{2}}$$

Just like writing in a journal can separate the wild and irrational feelings from the rational ones, we can take $\sqrt{18}$ and rewrite it as $3\sqrt{2}$. Feels good, doesn't it?

Here's the rule that lets us do this sort of thing:

What's It Called?

Product Property of Square Roots: For all positive (and zero) real numbers a and b:

$$\sqrt{ab} = \sqrt{a} \cdot \sqrt{b}$$

Notice how we can use this rule to write $\sqrt{36} = \sqrt{4 \cdot 9} = \sqrt{4} \cdot \sqrt{9} = (2)(3) = \mathbf{6}$. And we already knew that $\sqrt{36} = \mathbf{6}$, so we can see that this rule really works. *It allows us to find perfect squares and factor 'em out.* Radical expressions are considered to be in **simplified form** when *all* the perfect squares have been factored out.

Prime Factorization

To be totally sure we've found all the perfect squares, we can write out the radicand's *prime factorization,** and look for repeating prime factors. For example, if we want to simplify $\sqrt{180}$, we can write $\sqrt{180} = \sqrt{18 \cdot 10} = \sqrt{2 \cdot 9 \cdot 2 \cdot 5} = \sqrt{\underline{2 \cdot 2} \cdot \underline{3 \cdot 3} \cdot 5}$. Now we can see that two factors happen twice, 2 and 3, so we can separate them out:

$$\sqrt{2 \cdot 2} \cdot \sqrt{3 \cdot 3} \cdot \sqrt{5} = (2)(3)\sqrt{5} = \mathbf{6\sqrt{5}}$$

Personally, I like to search for the perfect squares themselves:

$$\sqrt{180} = \sqrt{18 \cdot 10} = \sqrt{2 \cdot 9 \cdot 2 \cdot 5} = \sqrt{4} \cdot \sqrt{9} \cdot \sqrt{5} = (2)(3)\sqrt{5} = \mathbf{6\sqrt{5}}$$

Do whichever you prefer!

Doing the Math

Put the following radical expressions in simplified form. I'll do the first one for you.

. . . .

* This just means to write a number as a product of its prime factors. To review this, check out Chapter 1 in *Math Doesn't Suck!*

1. $\sqrt{196} + \sqrt{117}$

<u>Working out the solution</u>: First, let's factor 196. Hunting for perfect squares, I always look for 4 and 9 first. Does 4 divide into 196 evenly? Yep! Try it, and you'll see that in fact $196 \div 4 = 49$. Yippee! And $49 = 7 \cdot 7$. So, $\sqrt{196} = \sqrt{4 \cdot 49} = \sqrt{4} \cdot \sqrt{49} = (2)(7) = \mathbf{14}$. For the second radical, hmm, 117 sure looks prime, but a divisibility trick,[*] $(1 + 1 + 7 = 9)$, tells us it's divisible by 9. In fact, $117 = 9 \cdot 13$. So, $\sqrt{117} = \sqrt{9 \cdot 13} = \sqrt{9} \cdot \sqrt{13} = \mathbf{3\sqrt{13}}$.

Answer: $14 + 3\sqrt{13}$

2. $\sqrt{50}$

3. $\sqrt{84}$

4. $\sqrt{108} + \sqrt{225}$

5. $\sqrt{324} + \sqrt{68}$

(Answers on pp. 409–10)

Supportive Friends . . . and Multiplying Radicals

As you know, some friends (even ones who seem nice!) can be a bad influence, encouraging us to ignore schoolwork, show up late to class, or stay up all night chatting instead of sleeping. However, truly supportive friends *want* to see us succeed and can bring out the best in us.

Radicals that are multiplied can bring out the best in each other, too. Notice that we can use that same rule on p. 284 to *multiply* radical expressions. After all, $\sqrt{ab} = \sqrt{a} \cdot \sqrt{b}$ is the same statement as

$$\sqrt{a} \cdot \sqrt{b} = \sqrt{ab}$$

For example, $\sqrt{3} \cdot \sqrt{5} = \sqrt{3 \cdot 5} = \sqrt{15}$.

. . . .

[*] This trick only works for 3 and 9. For more tricks, see p. 9 in *Math Doesn't Suck* or look online at danicamckellar.com/mds/extras.

The great thing about multiplying radicals together is that even if each of the radicals doesn't have its own perfect square, after you multiply them together, you might discover some *new* perfect squares if the two radicands had common factors.

See, on its own, there's nothing we can do with $\sqrt{6}$; it just doesn't have any perfect squares as factors, so we can't pull anything rational out, right? Same with $\sqrt{2}$. But look what happens when we multiply them together:

$$\sqrt{6} \cdot \sqrt{2} = \sqrt{6 \cdot 2} = \sqrt{12} = \sqrt{4 \cdot 3} = \sqrt{4} \cdot \sqrt{3} = 2\sqrt{3}$$

Separately, they remained totally irrational. Together, we were able to find some rational stuff to pull out. It looks like $\sqrt{6}$ and $\sqrt{2}$ are a good influence on each other!

QUICK NOTE   When we have radicals with coefficients stuck to their outsides, it's helpful to write the multiplication with dots and then rearrange factors to bring the radicals together.

$$3\sqrt{2} \cdot 4\sqrt{5} = 3 \cdot \sqrt{2} \cdot 4 \cdot \sqrt{5}$$

(rearranging the order of multiplication)

$$= 3 \cdot 4 \cdot \sqrt{2} \cdot \sqrt{5} = 12\sqrt{2 \cdot 5} = 12\sqrt{10}$$

Basically, we just collect all the radical stuff together and the non–radical stuff together and then multiply. Easy, right?

Watch Out!

There is no rule like this for addition or subtraction. So, $\sqrt{6} + \sqrt{2} \neq \sqrt{8}$, and $\sqrt{6} - \sqrt{2} \neq \sqrt{4}$. The truth is, there's no way to further simplify these; $\sqrt{6} + \sqrt{2}$ and $\sqrt{6} - \sqrt{2}$ just have to stay the way they are.

Doing the Math

Simplify the following expressions. I'll do the first one for you.

1. $\sqrt{50} \cdot 3\sqrt{14}$

<u>Working out the solution</u>: Because $\sqrt{50}$ has a perfect square inside (25), it can be rewritten as $\sqrt{50} = \sqrt{25 \cdot 2} = \sqrt{25} \cdot \sqrt{2} = \mathbf{5\sqrt{2}}$, which will make life easier when we multiply. Our problem is now $5\sqrt{2} \cdot 3\sqrt{14}$, which we can rearrange into $5 \cdot 3 \cdot \sqrt{2} \cdot \sqrt{14}$. We multiply to get $15\sqrt{2 \cdot 14} = 15\sqrt{28}$. Since $28 = \mathbf{4} \cdot 7$, let's factor out the new perfect square: $15\sqrt{4 \cdot 7} = 15 \cdot \sqrt{4} \cdot \sqrt{7} = 15 \cdot 2 \cdot \sqrt{7} = \mathbf{30\sqrt{7}}$. Done!

Answer: $30\sqrt{7}$

2. $\sqrt{18} \cdot \sqrt{2}$

3. $2\sqrt{30} \cdot 4\sqrt{5}$

4. $2\sqrt{21} \cdot 5\sqrt{7}$ *(Hint: Write 21 as 3 · 7 and leave it like that when you multiply.)*

(Answers on p. 410)

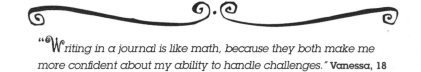

"Writing in a journal is like math, because they both make me more confident about my ability to handle challenges." **Vanessa, 18**

Combining Similar Radicals

As you know, we're allowed to add $2x + 3x = 5x$ because the variable parts are the same. However, there's nothing we can do to simplify $2x + 3y$.

Much in the same way, there's no way to combine $2\sqrt{6} + 3\sqrt{7}$ because the radical parts are different. But if the radical parts are the same, like $2\sqrt{7} + 3\sqrt{7}$, then we have *similar radicals*, and we can combine them exactly as if the radicals were x's:

$$2\sqrt{7} + 3\sqrt{7} = 5\sqrt{7}$$

Just think of $\sqrt{7}$ (or any irrational number, like π) as a variable or a box or a flower, and you'll be fine: To add/subtract similar radicals, just add/subtract their coefficients.

Watch Out!

I'm serious; when you have two radical expressions whose radical parts are identical, really and truly pretend they're variables. It's too easy to make a mistake like this: $\sqrt{3} + \sqrt{3} \rightarrow \sqrt{6}$. That is wrong! The correct answer is $\sqrt{3} + \sqrt{3} = 2\sqrt{3}$. It can be easier to see if we mentally substitute x for $\sqrt{3}$, knowing that $x + x = 2x$.

Step By Step

Simplifying radical expressions:

Step 1. Always obey the PEMDAS order of operations.* Note that this might alter the order of Steps 3 and 4, depending on the placement of any parentheses.

Step 2. If there are big numbers under the square root signs, look for perfect square factors to factor out. It'll prevent really big numbers from happening.

. . . .

* See p. 400 for a review of PEMDAS.

Step 3. For any multiplication: Expand it out with dots, and rearrange the factors to collect all the square root factors together and all the non-square root factors together. The square root factors will multiply together using this property: $\sqrt{a} \cdot \sqrt{b} = \sqrt{ab}$. Have any perfect squares emerged? If so, factor 'em out.

Step 4. For any addition/subtraction, only combine terms with *identical* radical parts. Imagine the radical part is a flower or a variable when you add/subtract them.

And... Action! Step By Step In Action

Simplify $-2\sqrt{15}(\sqrt{3} - \sqrt{75})$.

Steps 1 and 2. If wanted, we could use the distributive property, but we'd end up with really big numbers under the radical signs. So let's see if we can simplify the inside of the parentheses first. We'll rewrite $\sqrt{75} = \sqrt{25 \cdot 3} = \sqrt{25} \cdot \sqrt{3} = 5\sqrt{3}$.

Steps 3 and 4. Well, lookie there; now the inside of the parentheses looks like this: $\sqrt{3} - 5\sqrt{3}$, which we can subtract because the radical parts are the same. And just as $x - 5x = -4x$, it's true that $\sqrt{3} - 5\sqrt{3} = -4\sqrt{3}$.

Our problem now looks like this: $-2\sqrt{15}(-4\sqrt{3})$. Let's expand the multiplication with dots: $(-2) \cdot \sqrt{15} \cdot (-4) \cdot \sqrt{3}$

(rearranging) $= (-2) \cdot (-4) \cdot \sqrt{15} \cdot \sqrt{3}$
$$= 8\sqrt{15 \cdot 3} = 8\sqrt{45}.$$

Oh hey, 9 is a factor of 45 *and* a perfect square, so we can pull it out: $8\sqrt{45} = 8\sqrt{9 \cdot 5} = 8 \cdot \sqrt{9} \cdot \sqrt{5} = 8 \cdot 3 \cdot \sqrt{5} = 24\sqrt{5}$. Pant, pant. Done!

Answer: $24\sqrt{5}$

Simplify these expressions. I'll do the first one for you.

1. $\sqrt{6} - 4\sqrt{6} \cdot \sqrt{4}$

<u>Working out the solution</u>: According to PEMDAS, we must do the multiplication first,[*] which is $4\sqrt{6} \cdot \sqrt{4} = 4\sqrt{6} \cdot 2 = 4 \cdot 2 \cdot \sqrt{6} = 8\sqrt{6}$. Okay, now the entire problem looks like this: $\sqrt{6} - 8\sqrt{6}$. These have the same variable part, so we can subtract them. And just as $x - 8x = -7x$, we can do $\sqrt{6} - 8\sqrt{6} = -7\sqrt{6}$. Done!

Answer: $-7\sqrt{6}$

2. $4\sqrt{5} + 3\sqrt{5}$

3. $2\sqrt{20} + \sqrt{45}$

4. $2\sqrt{20} - \sqrt{45}$

5. $\sqrt{45} - 2\sqrt{20}$

6. $4\sqrt{5} - 3\sqrt{5} \cdot \sqrt{4}$

(Answers on p. 410)

Takeaway Tips

Radical is another word for square root, cube root, etc.

If there are any perfect square factors under a square root sign, pull 'em out using this rule: $\sqrt{ab} = \sqrt{a} \cdot \sqrt{b}$.

To multiply radicals, just separate out all the multiplication with dots, rearrange factors, and remember that $\sqrt{a} \cdot \sqrt{b} = \sqrt{ab}$.

To add/subtract two radical expressions with identical radical parts, just add/subtract their coefficients exactly as you would if the radical parts were variables.

. . . .
[*] We'd get a totally wrong answer if we first subtracted $\sqrt{6} - 4\sqrt{6}$. Obey PEMDAS!

Confessional: How Can I Focus and Avoid Distractions?

A huge percentage of kids these days are being diagnosed with ADD or ADHD. But believe me when I say that everyone struggles with being able to focus on schoolwork, *everyone*.* Luckily, there are steps we can *all* take to focus and achieve our goals, whatever they might be—steps that have nothing to do with potentially addictive drugs! Whether it's to study for a test or to read two chapters of your English novel, read below and see what others are saying about what distracts them and how they have learned to focus.

"To help myself focus in class, I like to ask a lot of questions. It keeps me 'in' the class, if you know what I mean. Even if I already know the answer to something, I'll ask a question. It really works." **Monique, 16**

"I like to make 'beats' to my homework. I make up songs with the things I'm learning—it makes it easier to focus on homework and also easier to succeed on tests, because I can just sing the songs I made up to remember everything." **Shawnee, 16**

"Family problems are my biggest distraction. To do my schoolwork, I just use my imagination and 'minus' the drama out of my brain for a while. I pretend it doesn't exist. That works most of the time." **Tasha, 17**

"Ironically, bad grades distract me from my schoolwork, because I get discouraged and don't feel like working at all. But I deal with it with my motto, 'If you're at the bottom there's only one way to go—up.'" **Dustin, 15**

"I have ADHD so yes, everything is a distraction. But I can control it by willpower. I deal with distractions by putting extra effort into focusing, even when I don't feel like it." **Ben, 16**

* Yes, even me!

"Go to a different place!"

"I'll sit in my backyard alone, which is my favorite place to do homework." **Chelsea, 14**

"Video games are my biggest distraction. I have to go into a different room to study." **William, 15**

"It's always hard for me to work at home because we have a lot going on with my animals, the TV, and my dad. Normally I try to finish my work at school." **Carter, 14**

"Distracted by People"

"My biggest distraction? Probably my little brothers! How do I deal with them? Usually closing the door to my room is enough!" **Tori, 14**

"I usually say to my boyfriend, 'Look, I like spending time with you, but I need to focus on my work, and I'll see you on weekends. Okay?' It always works out." **Tabitha, 15**

"I think technology is the biggest thing that interferes with me focusing on my schoolwork. Texting, television, social networks, and instant messaging available 24/7 make it very easy to get distracted. If I make goals for myself, it makes it easier to focus. For example, I'll say to myself, 'As soon as I finish math and English, I can text so-and-so, but in 20 minutes I have to work on science.' That's how I deal with my distractions: by disciplining myself to only get distracted in small increments." **Gussie, 14**

"I was diagnosed with ADD, but it seems like a lot of kids are these days, so I'm not sure what that even means anymore. I think it's hard for everyone to focus, and it's a challenge we all share on some level." **McKenna, 15**

If you want to avoid distractions, sometimes the best way is to actually *avoid* them! Study in a different room from your computer and phone, or turn them off. That makes it more difficult to jump on the Internet "real fast," which of course ends up being much longer than you meant for it to be. We ALL struggle with this. Most of the time, I disconnect my Internet while I'm working, because I get distracted, too! It's just so easy to wonder about something like, "Hey, who sings that song again?" and then quickly look it up . . . and then get distracted by other stuff or suddenly remember something else you wanted to do . . . and then a quick 2 minutes becomes 2 hours or more.

Do what you have to in order to stop the cycle. Disconnect the Internet, turn off your phone, go into a different room with no TV or computer, study in the library, go outside—whatever works. When the distractions come from inside your own brain, like daydreaming or creative thoughts, say to them, "Why thank you so much, creative part of my brain! I'm grateful to have you, but right now isn't the best time. Let's think about this later." We don't want to beat ourselves up for daydreaming. Creative imagination is a wonderful gift! But when your creative mind is interfering with a goal, just write yourself a quick note about what you were thinking about so you can think about it all you want later on.

Then be sure to reward yourself when you're done with your assignments! And get plenty of sleep, too. You'll look and feel so much better with 8 to 9 hours of sleep, and you'll be much better able to focus the *next* day.

Believe me when I tell you, all of this will go a long way to help you in "real life" as an adult. Because as an adult, there are even more distractions and other responsibilities, like paying bills. So this is a great time to get good at beating the distractions and procrastination, and learn to focus. Everyone is a little different, so experiment to see what works for you!*

. . . .

* If you have a copy of *Math Doesn't Suck*, be sure to see the personality quiz on p. 110 and the Troubleshooting Guide on p. 265 for more ideas.

Chapter 20

Chocolates for Breakfast
Intro to Polynomials and Adding/Subtracting Them

I love chocolate, especially dark chocolate—though not usually for breakfast. I like to eat three healthy meals a day, filled with organic raw fruits and vegetables and plenty of protein. I stay away from greasy foods, sugar, and anything artificial or overly processed. Sometimes it's hard to keep my meals healthy, but it's always worth it! And then, of course, I reward myself with a little dark chocolate now and then.

Meals show up in math, too—sort of. We'll learn a bunch of math vocab in this chapter, and four of the new words end in *-mial*, pronounced "mee-al." Like if you were really hungry, you might say, "I need a mee-al!"

Yes, we'll learn words like mono<u>mial</u>, bino<u>mial</u>, trino<u>mial</u>, and polyno<u>mial</u>. You didn't know there were meals besides breakfast, lunch, and dinner, didja? Don't worry, I'll help you keep them all straight.

Ring Ring What's It Called?

A **monomial** is a single expression consisting of numbers and/or variables all stuck together with multiplication (or division),* but not addition or subtraction. Here's one: $5x^2$. *Monomials* sometimes have more than one variable, like $7x^3yz^6$.

. . . .

* In fraction form. BTW, some textbooks consider terms like $\frac{3}{x}$ to be monomials, but most would say variables have to have *positive* exponents to be considered a monomial. (In this case, x's exponent is −1.) The truth is, this distinction isn't very important because it just doesn't come up much!

These are each monomials: z 0.07 $\dfrac{11y^3}{2}$ $10ab^2c$ 0

These are *not* monomials: $5 + x^2$ $9a^2 - a^2$

Huh, so something like 0.07 is considered a monomial? That's a pretty small mee-al. More like a little dark chocolate treat, if you ask me.

Ring Ring What's It Called?

The **degree** of a monomial is the exponent on the variable, or if there's more than one variable, then the *degree* is the <u>sum</u> of the stuck-together variables' exponents.

$$\text{degree} = 10$$
$$\dfrac{1}{8}a^{10}$$

$$\text{degree} = 5+2 = 7$$
$$3x^5y^2$$

For example, the *degree* of $\frac{1}{8}a^{10}$ is 10, and the degree of $3x^5y^2$ is 7, because $5 + 2 = 7$. Also, the *degree* of $4x$ (in other words, $4x^1$) is 1.

By the way, the *degree* of any nonzero number, like $\frac{1}{2}$ or –9, is just 0. (See the QUICK NOTE on the next page.)

Since the exponents look like they're way up in the sky, checking the *degree* is like checking the weather outside. "Hey, what's the weather like up in the sky? How many degrees is it up there?" And then you look at the exponent(s). Voilà!

Watch Out!

When adding variables' exponents to determine the degree, don't forget the invisible exponents of 1. For example, the monomial xy^2z has a degree of 4, because x and z both have exponents of 1, and $1 + 2 + 1 = 4$.

QUICK NOTE   So, nonzero constants like 5 and −8 have a degree of 0. (Their exponents are 1 but their degrees are 0!) But what about zero itself? It's said to have "no degree." Yes, "no degree" is somehow even a lower status than having a degree of 0. Geez, that's rough.

Like Monomials

Like, omigod, monomials. Just kidding. Two monomials are considered **like** if their variable parts are identical. So the following are all *like monomials*:

$$0.9a^2b \qquad ba^2 \qquad (-23)a^2b \qquad \frac{4}{5}ba^2$$

See? Their variable parts are all the same. Notice that the *order* of the variables doesn't matter. By the way, the degree of each of these monomials is 3, because $2 + 1 = 3$. On the other hand, the monomials below are all *unlike* because none of their variable parts exactly match:

$$4x^2y^3 \qquad 4x^2y^2 \qquad 4x^2 \qquad 4w^2y^3$$

The thing about *like monomials* is that we can combine them; for example, we can add $ba^2 + (-23)a^2b = -22a^2b$. Since they have the same variable parts, we just combine their coefficients, and we're done! So, how does all of this relate to *polynomials*? I know you're just breathless with anticipation.

What Are They Called?

A **polynomial** is a string of monomials that are connected by addition and subtraction. The monomials are called the **terms*** of the polynomial. For example, the polynomial below has five terms:

$$3xy^4 - 2xy^4 - 9x^2y + 6x^2 - 7$$

* For more on *terms*, see p. 3.

If none of its monomial terms are *like*, then the polynomial is in **simple form**. For example, the above polynomial is not in simple form, because its first two terms have the same variable part. But if we combine the first two terms, $3xy^4 - 2xy^4 = \mathbf{xy^4}$, then it is in *simple form* (and now the polynomial has four terms).

$$xy^4 - 9x^2y + 6x^2 - 7$$

To determine the **degree of a polynomial**, make sure it's in simple form, and then look for the *highest* (biggest) degree on all of its terms. For this one, look up, check the "weather," and notice that the degrees of its terms are, in order: 5, 3, 2, 0. So, the degree of the whole polynomial is just the greatest of those, which is 5.

QUICK NOTE   Generally speaking, it's best to write polynomials' terms in order from greatest degree to least degree. Because the degrees get smaller, left to right, this is called *descending order*. It's very popular with teachers. Also, if the polynomial has a lot of terms, it makes it easier to tell if there are any *like terms* left that need to be combined.

Just when you thought you were done with the vocabulary, there's more! Believe it or not, polynomials of different lengths have different names. Yep. So a polynomial with just two terms is called a **binomial**, and a polynomial with three terms is called a **trinomial**. Oh, why do they insist on making life more complicated than it needs to be? I can't answer that question, but I *can* make it easier to remember the names.

What Are They Called?

Monomial, Binomial, Trinomial, Polynomial

Check out these prefixes in other contexts; it makes it easier to keep it all straight.

<u>Mono:</u> *one*

<u>Mono</u>cle—an eyeglass for just *one* eye, good for creepy villains

<u>Mono</u>logue—an acting scene for just *one* person

<u>Mono</u>mial—has *one* term, for example, $3x$ or a^2b. A small, single "mee-al," like maybe a banana with peanut butter smeared on it. Yummy breakfast!

<u>Bi:</u> *two*

<u>Bi</u>cycle—has *two* wheels

<u>Bi</u>lingual—speaks *two* languages, very useful!

<u>Bi</u>nomial—has *two* terms, for example, $2a - 1$ or $x + y$. This "mee-al" would consist of two parts, like a soup and salad combo. Good for lunch.

<u>Tri:</u> *three*

<u>Tri</u>logy—has *three* parts

<u>Tri</u>plets—*three* babies born together

<u>Tri</u>nomial—has *three* terms, for example, $x - 3y - 5$ or $2x^2 + 5x + 1$. This "mee-al" has three parts, like chicken with a side of carrots and a baked potato. Dinner, perhaps?

<u>Poly:</u> *more than one, many*

<u>Poly</u>gamy—married to *more than one* person at a time. Perhaps *many*. Yikes!

<u>Poly</u>nesia—a group of *many* islands (more than 1000!) scattered across the Pacific Ocean, including Samoa, Tonga, Hawaii, and New Zealand.

<u>Poly</u>nomial—often has *more than one* term, perhaps *many*: $3c + d$ or $4x + x^2y - 5z + 8$. Now we're talking Thanksgiving: turkey, salad, mashed potatoes, cranberry sauce . . . mmm. (Note: monomials, binomials, and trinomials are all specific types of polynomials.)

Let's practice keeping all this stuff straight!

Doing the Math

For each polynomial, do the following:

a. How many terms does it *currently* have?

b. Write it in simple form (combine like terms), and put it in descending order (of degrees).

c. Now that it's in simple form, what is its degree, how many terms does it have, and is it a monomial, a binomial, or a trinomial?

I'll do the first one for you.

1. $x^7y^2 - 8s^2 - s^2t + 3s^2$

<u>Working out the solution</u>: It's currently made up of **four terms**. Done with part a! For part b, there are two *like* terms, $-8s^2$ and $3s^2$, so we can combine them: $-8s^2 + 3s^2 = -5s^2$. We're now ready to write it in descending order: $x^7y^2 - s^2t - 5s^2$. For part c, the highest degree for any term is $7 + 2 = 9$, so that means the polynomial has a degree of 9. Now it has just three terms in it, which means it's a trinomial. Done!

Answer: a. four terms; b. $x^7y^2 - s^2t - 5s^2$; c. degree 9; three terms; trinomial

2. $wz^2 - 1 + 5wz^2$

3. $3x + 9 - 3x - 1$

4. $e^4 + 5 - d^2e^3 - d^2e^3 + 1$

(Answers on p. 410)

Factoring the GCF out of Polynomials

In Chapter 3, we factored out the greatest common factor (GCF) by using
reverse distribution and the Birthday Cake method, which is hands-down
the easiest way to factor polynomials. It's the exact same method but
with bigger exponents now.

QUICK NOTE   Just like in Chapter 3, we'll check our
answer by multiplying it out to make sure we get what
we started with. But this time, we'll have higher
powers, so we'll end up adding exponents[*] on similar
bases (variables).

Doing the Math

Factor each polynomial using the Birthday Cake method. I'll do
the first one for you.

1. $5b^8c^8 - 4b^3c^6d + 8b^3c^5$

. . . .

* See p. 241 to review. In a nutshell, when we multiply two bases that are the same,
we just add their exponents: $b^m \cdot b^n = b^{m+n}$.

<u>Working out the solution</u>: They're just exponents—they're not going to bite us, promise! Let's set up the birthday cake. First, notice that the *coefficients* share no common factors, so there will be no numbers to pull out. Hmm, all terms seem to have powers of *b*, and the lowest of those powers is b^3, so we write it on the side and factor it out, remembering to bring down all the + and − signs. So $5b^5c^8 - 4c^6d + 8c^5$ is our next layer. Next, it seems that powers of *c* are in every term, and the smallest is c^5. Factoring out c^5, we get $5b^5c^3 - 4cd + 8$, which has no common factors to all of its terms so it's the last layer.

$$b^3 \overline{\left| \, 5b^8c^8 - 4b^3c^6d + 8b^3c^5 \right.}$$
$$5b^5c^8 - 4c^6d + 8c^5$$

$\rightarrow$

$$b^3 \overline{\left| \, 5b^8c^8 - 4b^3c^6d + 8b^3c^5 \right.}$$
$$c^5 \overline{\left| \, 5b^5c^8 - 4c^6d + 8c^5 \right.}$$
$$5b^5c^3 - 4cd + 8$$

Just like in Chapter 3, we now multiply the left side for our GCF, b^3c^5, and the bottom layer goes in parentheses, so our answer is $b^3c^5(5b^5c^3 - 4cd + 8)$. Let's check our work by multiplying out our answer, mentally adding *b*'s exponents, $3 + 5 = 8$, and *c*'s exponents, $5 + 3 = 8$ and $5 + 1 = 6$. We get the original polynomial:

$b^3c^5(5b^5c^3 - 4cd + 8) = 5b^8c^8 - 4b^3c^6d + 8b^3c^5$. Yep!

Answer: $b^3c^5(5b^5c^3 - 4cd + 8)$

2. $2a^5b^4 + 6a^4b^5 + 9b^4a^5c$

3. $3z^8 - 12y^2z^5 + 18y^3z^4 - 6z^3$

4. $14gh^6k^2 - 42kg^4h^7 + 49h^6k^4g$

(Answers on p. 410)

Adding and Subtracting Polynomials*

Time for doodling. Yes, when teachers spring long polynomials on us, there is no better way to keep like terms straight when adding and subtracting. Pick a different, creative underline style for each variable type, doodle, and then proceed.

Hey, the next time your teacher catches you doodling during class, you could say, "But I was just practicing my polynomials." Okay, I guess it depends on the doodle.

QUICK NOTE   We can use the word *monomial* or *term* interchangeably in a polynomial. In this context, they mean the same thing.

Step By Step

Adding/subtracting polynomials:

Step 1. If the polynomials are being *added*, then drop the big parentheses, if any. If the polynomials are being *subtracted*, then first distribute the negative sign.

Step 2. I highly recommend writing all remaining subtraction as "adding a negative," and if any coefficients aren't written in front of variables, write them in as 1 or −1.

Step 3. Give each type of *like term* its own underline style.

Step 4. Combine like terms (*like monomials*) by combining their coefficients.

Step 5. Leave the answer in descending order, from highest degree to lowest, and streamline† your negatives, if any. Done!

. . . .

* This is just a more advanced version of what we did in Chapter 9 of *Kiss My Math*. So if any of this is confusing, you know where to go.
† For more on streamlining negatives, see p. 6.

QUICK NOTE   The hardest part of this stuff is keeping track of everything. So don't skip steps, and have fun with your creative underlines. It'll keep you alert and focused.

Doing the Math

Add or subtract these polynomials, and leave them in descending order. I'll do the first one for you.

1. $(3y^2x - 2 + 4y^2) - (3xy^2 - y^2 + 6)$

<u>Working out the solution</u>: Distributing the negative sign, we'll <u>change the sign</u> of each term in the second set of parentheses: $3y^2x - 2 + 4y^2 - 3xy^2 + y^2 - 6$. Now, let's write each subtraction as "adding a negative." Let's also write in the 1 and -1 coefficients, and underline like terms. (And we'll rewrite the first term, $3y^2x$, as $3xy^2$.) Doodle time!

$$3xy^2 + (-2) + 4y^2 + (-3xy^2) + 1y^2 + (-6)$$

First, let's combine the <u>heart</u> terms: $3xy^2 + (-3xy^2) = \mathbf{0}$. How nice! Next, let's combine the <u>dotted</u> terms, $(-2) + (-6) = \mathbf{-8}$, and the <u>double wavy</u> terms, $4y^2 + 1y^2 = \mathbf{5y^2}$. So we have $\mathbf{0 + (-8) + (5y^2)}$. Putting it in descending order, we get $(5y^2) + (-8)$. Let's streamline the negative for the final answer. Done!

Answer: $5y^2 - 8$

2. $(a^2 - b^2) - (a^2 + 2b^2)$

3. $(g^2h + 4gh) - (6gh - gh^2)$

4. $(z^2 + 5z - 2y^3 - 3) + (2 - z^2 + 3y^3 - 5z)$

5. $(2x^6 - 7 + 2x^4 - 3x^5) - (x^6 - 3x^3 + 2x^5 + 4)$

(Answers on p. 410)

Takeaway Tips

 The degree of a polynomial is the <u>highest</u> degree of any of its monomials, and the degree of each monomial is the exponent of its variable or the <u>sum</u> of the variables' exponents. Don't forget to add the invisible 1's.

To factor out the GCF from a polynomial, use the Birthday Cake method, and then check your work by distributing. The rule for adding exponents comes in handy for this.

When combining like terms, use creative underlines to keep organized. Doodle!

It's best to leave polynomials in simple form, meaning that all *like terms* have been combined, and in descending order: from the highest degree monomial to the least.

Body Language

Some students just don't like raising their hands in class, which can be a sign of shyness, insecurity, or low self-esteem. You can probably tell the shy kids from the confident ones from that evidence alone!

But did you know that we can also learn a lot about people just by noticing their posture? It's called "body *language*" because through posture alone, we are actually communicating to the world around us, telling people who we are. The next time you think of it, try watching your friends—even perfect strangers—and notice the way they walk and the way they sit in a chair. Are they confident? Do they hold their head up and shoulders back? Or are they hunched over, almost as if they're hiding from the world? What is their body language *telling* you?

From a young age, girls are taught to be polite. Think about how most of the boys you know sit, and compare that with the girls in your class. Boys tend to sit with wide legs, place their arms on the backs of the chairs next to them, speak in a big full voice, and take up a lot of space. Girls, on the other hand, often hunch over, cross their legs, draw their shoulders forward, keep their hands in their laps, speak in a soft voice, and generally speaking, make themselves "smaller" than they are. This is the body language version of "polite": not taking up too much *space*.

I mean, there's a time and a place for everything, and no one wants to be an obnoxious, gum-smacking gal sitting wide-legged with a skirt on—oops! But still, there's a big difference between being rude and being unapologetically confident.

Look around, and you'll notice that the young women who seem truly confident will stand up straight, smile, and speak in a full, grounded voice. And these women could be skinny, overweight, beautiful, or plain. It doesn't matter; they automatically look more attractive because of their attitude, and it shows in their body language without them having to say a word.

Have you thought about what *your* body language is communicating?

If you're thinking, "I have insecurities, so I end up hunched over. I can't help it," then listen up: *Everyone* has insecurities—yes, me too—but we can train ourselves to focus on the positive. By practicing good posture, you'll not only look more confident, you'll *feel* more confident. Take a chance, and let yourself be bold. Raise your hand in class, make a new friend, and walk with your head up and shoulders back.

Take it from me: Not only can good posture prevent back pain later in life, but it will help you discover your most confident self today!

TESTIMONIAL

Sarah Huck
(Racine, WI)
Before: Didn't think she needed math
After: Professional cook and recipe developer!

I was pretty good at math as a kid. But when I got to high school in ninth grade, my math grades started slipping—fast.

I convinced myself that I wasn't cut out for more advanced math, but deep down, I knew the real reason I wasn't doing well anymore. I had decided that math required too much work and focus. All my friends goofed off, passed notes, and talked to cute boys during class. I figured, why not me, too? It was easy to get lazy in math because it seemed really unimportant, and I couldn't imagine that I would ever need it. I mean, I wanted to be a chef. I didn't need math, right?

"When you cook as much as I do, you really can't avoid math."

When I dreamt of being a professional chef and recipe developer, I only imagined concocting gorgeous, three-tiered chocolate cakes or beautifully plated salmon fillets. When I found out how much math was used, I was terrified! I was so afraid that I'd miscalculate something and make a huge mistake that would make me look foolish and cost me my job. And guess what? I almost did.

My first year, I created recipes for a writer who worked for a big, fancy food publication. I was so proud to send the writer one of my favorite recipes, but I didn't realize how precise my measurements and conversions had to be. Well, it turns out that she sent the recipe to her editor, who then gave it to a recipe tester to double-check the work. And lo and behold, the recipe was a disaster! Everyone was furious, and I was absolutely mortified. It really, really taught me that I had to take the math I was doing seriously, and I was so grateful when my boss gave me another chance.

From that point on, I had to admit when I didn't know how to do things, and I asked tons of questions. I hated having to do it, but I figured it was better than the alternative—messing up again! With years of practice, I've been able to build my math skills to what they could have been all along, if I'd only paid more attention in class. And believe me, I use my math skills many times a day.

Sometimes I need to convert measurements like grams (used in professional kitchens) to ounces (used in homes). I also convert weights to volume measurements. Other times, I have to scale recipes to adjust the number of servings, like taking a recipe for four people and adjusting it for 50 people, or vice versa. This can be extra challenging when not all of the ingredients are easily divided to accommodate standard kitchen measurements, like dividing $\frac{1}{3}$ cup in two or dividing two eggs into thirds! My kitchen "math kit" includes a scale, three kinds of thermometers (meat, candy, deep-fat), a timer, a ruler, teaspoons, liquid cup measures, and dry cup measures. When you cook as much as I do, you really can't avoid math.

To be honest, one of the things I've always loved most about working in food is that it is a chance to exercise all kinds of subjects at once—math, science, history, literature, politics, and of course, art (the food version!). I never get bored. Plus, unlike most people out there in the working world, I get to eat all of my projects at the end of the day!

Chapter 21

A Day at the Hair Salon

Multiplying Polynomials and the FOIL Method

*A*h, the many uses of foil. Kindergarten teachers use gold-foil stars for A+ papers, leftovers often leave restaurants in foil, and I like to mix paint colors on foil. It makes a great art palette! Some people even make little *hats* out of foil, because they believe it will protect them from evil alien gamma rays.*

From time to time, I go to the hair salon to get highlights, and they wrap little hair segments with special hair bleach in—you guessed it: foil. The bleach is really toxic smelling though, so I end up sitting there, holding my breath, with foil all over my head. Where's the camera when you need it?

Well, "foil" is about to help us with something else: simplifying complicated-looking algebraic expressions, which can seem pretty alien, too. We know how to use the distributive property to multiply things like $x(3y + z)$; as in Chapter 3, we'd just imagine that x is arriving to a party, and she says "hi" to both the $3y$ and the z. We'd get $3xy + xz$. But how do we multiply two *binomials* together?

$$(x + 2)(3y - z)$$

As it turns out, there's a really nice method that the ladies down at the hair salon like to call . . . FOIL.

. . . .

* I'm pretty sure most people know the tin-foil hats are just a silly joke. And let's face it, if aliens (evil or not) were really trying to send us messages, would a tin-foil hat really keep them out?

Shortcut Alert

FOIL stands for "**F**irst, **O**utside, **I**nside, **L**ast." It describes an easy way to multiply two binomials together. For any real numbers *a, b, c,* and *d*:

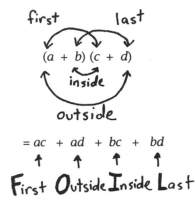

$$= ac + ad + bc + bd$$

We multiply the **First** two terms together, then we multiply the **Outside** terms together, then we multiply the **Inside** terms together, and finally we multiply the **Last** terms together. We add 'em all together, and then we have successfully multiplied the two binomials. Say it with me: First, Outside, Inside, Last—FOIL!

When you think about it, all FOIL is really doing is allowing both girls to take turns saying hello. So *a* gets to say hi to everyone inside the party, and then *b* takes her turn. Notice that they don't have to say hi to each other; after all, they've probably already met, seeing as how they arrived in the same set of parentheses.

Step By Step

Multiplying binomials with FOIL:

Step 1. Multiply the **F**irst two terms together, then the **O**utside terms, then the **I**nside terms, and finally the **L**ast terms. Don't forget the negative signs!

Step 2. Add the products together, and simplify by adding like terms, if any.

Step 3. Make sure to leave the polynomial in descending order,* and streamline any negative signs into subtraction. Done!

QUICK NOTE   We'll use lots of new words like binomial and *trinomial* that we learned in Chapter 20, so you can always look back if you can't remember one of them.

Multiply $(3x^2 + 5x)(x - 2)$.

Step 1. Time for FOIL!

<u>First</u>: We multiply the first terms together:

$$(3x^2 + 5x)(x - 2) \quad \text{first:} \ 3x^2 \cdot x = \underline{3x^3}$$

<u>Outside:</u> Next we multiply the outside terms together (keeping that negative!):

$$(3x^2 + 5x)(x - 2) \quad \text{outside:} \ 3x^2 \cdot (-2) = -\underline{6x^2}$$

<u>Inside</u>: Then we multiply the inside terms together:

$$(3x^2 + 5x)(x - 2) \quad \text{inside:} \ 5x \cdot x = \underline{5x^2}$$

<u>Last</u>: Finally, we multiply the last terms together (again, notice the negative sign):

$$(3x^2 + 5x)(x - 2) \quad \text{last:} \ 5x \cdot (-2) = \underline{-10x}$$

. . . .

* See p. 297 for an explanation of descending order.

Step 2. And now, we just add them together: $3x^3 + (-6x^2) + 5x^2 + (-10x)$.

In this case, the two middle terms are *like terms*, and we can combine them: $(-6x^2) + 5x^2 = -x^2$. Now our answer looks like this:

$$3x^3 + (-x^2) + (-10x)$$

Step 3. It's already in descending order, but we should streamline the negative signs into subtraction. Done!

Answer: $3x^3 - x^2 - 10x$

QUICK NOTE   Let's face it; it's just not practical to draw the arching arrows each time. Instead, use your pencil to "tap" the two terms that you'll be multiplying, and use scratch paper if you want to keep your homework neat. So in this example, you'd first tap the $3x^2$ and the x with your pencil, thinking to yourself "first," and you'd write $3x^2 \cdot x = \underline{3x^3}$ on your scratch paper. Next, while thinking to yourself "outside," you'd tap the $3x^2$ and the -2, and so on, making sure to keep the negative signs. It's always worked for me, anyway!

Watch Out!

Pay attention to negative signs; they need to come along for the ride! When we multiplied the "outside" terms of $(3x^2 + 5x)(x - 2)$, it would have been easy to get: $3x^2 \cdot 2 = 6x^2$ instead of $-6x^2$. Oops! And then when we added that to $5x^2$, we'd get $11x^2$, and we'd be looking at our wrong answer thinking, *What did we mess up?*

QUICK NOTE   In a problem like $7 \cdot 3 = 21$, the **factors** are 7 and 3, and their **product** is 21. Well, for the multiplication problem $(3x^2 + 5x)(x - 2) = 3x^3 - x^2 - 10x$, the *factors* are $(3x^2 + 5x)$ and $(x - 2)$, and their *product* is $3x^3 - x^2 - 10x$. This is a great time for you to start thinking of the parts of all multiplication problems as *factors* and *products*.

Doing the Math

Use FOIL to multiply these binomial factors. Leave their product in simple form (combine like terms) and in descending order. I'll do the first one for you.

1. $(x + 2)(3y^2 - z)$

<u>Working out the solution</u>: First, $x \cdot 3y^2 = $ **$3xy^2$**; Outside: $x \cdot (-z) = $ **$-xz$**; Inside: $2 \cdot 3y^2 = $ **$6y^2$**; Last: $2 \cdot (-z) = $ **$-2z$**. Adding them together, we get **$3xy^2 + (-xz) + 6y^2 + (-2z)$**. We can flip the positions of the second and third terms; <u>either way</u>, we'll be in descending order. And since there are no like terms to combine, we just streamline the negatives and we're done!

Answer: $3xy^2 + 6y^2 - xz - 2z$

2. $(2x + 1)(x - 3)$

3. $(z + 2)(w^2 - 3)$

4. $(a + b)(a - b)$

(Answers on p. 410)

Specialty FOIL—Squares!

This isn't as much like origami* as it sounds. Nope, in fact we're going to take binomials and *square* them. Let's look at $(x - 3)^2$. Hmm, we're not allowed to distribute that exponent over subtraction, so what can we do with it? When in doubt, multiply it out!

$$(x - 3)^2 = (x - 3)(x - 3)$$

And now we can just apply FOIL, like we would any other time.

First: $x \cdot x = \mathbf{x^2}$

Outside: $x \cdot (-3) = \mathbf{-3x}$

Inside: $-3 \cdot x = \mathbf{-3x}$

Last: $-3 \cdot (-3) = \mathbf{9}$

area =
$(x-3)^2$ $\Big\}$ $(x-3)$

$(x-3)$

Combining like terms, we get $x^2 - 3x - 3x + 9 = \mathbf{x^2 - 6x + 9}$. Even though it doesn't look like it, we know that this is a perfect square; in fact, it's a **perfect square trinomial**.[†]

Ring Ring What's It Called?

A **perfect square trinomial** is the *product* we get from squaring a binomial.

$\mathbf{x^2 - 6x + 9}$ is a *perfect square trinomial*, because:

$$(x - 3)^2 = x^2 - 6x + 9$$

$\mathbf{y^2 + 8y + 16}$ is a *perfect square trinomial*, because:

$$(y + 4)^2 = y^2 + 8y + 16$$

Notice that in perfect square trinomials, the first and last terms will always be *perfect squares*. Also, the "inside" and "outside" terms will always be the same as each other, and then we combine them, since they are like terms. It's always a good idea to keep an eye out for patterns!

. . . .

* The Japanese art of paper folding, where foil squares become miniature birds or frogs.
† *Trinomial* just means it has three terms (see p. 298).

Doing the Math

Evaluate the following squared binomials by first writing them out as two factors multiplied together and then using the FOIL method. I'll do the first one for you.

1. $(5h - 6)^2$

<u>Working out the solution</u>: What does it mean to square this binomial? It means that we *multiply it times itself*, so $(5h - 6)^2 = (5h - 6)(5h - 6)$. Applying FOIL with a tap of the pencil: The <u>first</u> terms multiply to give $5h \cdot 5h = \mathbf{25h^2}$, <u>outside</u> terms $5h \cdot (-6) = \mathbf{-30h}$, <u>inside</u> terms $(-6) \cdot 5h = \mathbf{-30h}$, and <u>last</u> terms: $-6 \cdot (-6) = \mathbf{36}$. Adding them together, we get $25h^2 + (-30h) + (-30h) + 36$. Combining the middle two terms, we get $25h^2 + (-60h) + 36$. And streamlining the negative sign, we get our answer.

Answer: $25h^2 - 60h + 36$

2. $(x - 1)^2$

3. $(3y - 2)^2$

4. $(a + b)^2$

5. $(a - b)^2$

(Answers on p. 410)

Beyond FOIL: Multiplying Big Polynomials, and General Party Etiquette

FOIL only works when we're multiplying two binomials. So how about something like $(2 + a)(3 + 2a - b)$ or even $(a + b + c)(d + e + f)$? No problem!

Time to go back to our party. When a group of girls arrives at a party, I think you'll agree it's polite for each one to individually say hello to everyone who is already at the party, no matter how many guests are there. As you know, they don't need to say hi to each other because they arrived together. Here's what happens:

$$(a + b + c)(d + e + f) = ?$$

For a's "hellos," imagine an arrow going from a to each term in the second parentheses. We get $ad + ae + af$. For b's "hellos," imagine an arrow going from b to each term in the second parentheses, and we get $bd + be + bf$. For c's "hellos," imagine an arrow going from c to each term in the second parentheses, and we get $cd + ce + cf$. Then for the answer, we add 'em up:

$$(a + b + c)(d + e + f) = ad + ae + af + bd + be + bf + cd + ce + cf$$

That's a lot of hellos. Yeah, that looks pretty long and scary, but you know what? We did it! It's all about staying organized and remembering to be polite. By the way, there was no way to further simplify this particular answer. If you look, you'll see that none of these terms are *like terms*, so there's nothing to combine.

All party talk aside, here's the formal way to describe multiplying polynomials:

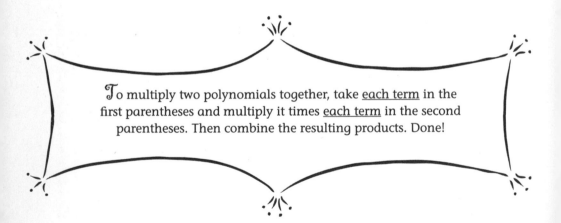

To multiply two polynomials together, take each term in the first parentheses and multiply it times each term in the second parentheses. Then combine the resulting products. Done!

 Doing the Math

Multiply these polynomials, and leave them in simple form and in descending order. Remember to say hi to everyone at the party! I'll do the first one for you.

1. $(a - b)(a^2 + ab + b^2)$

<u>Working out the solution:</u> First, a says hello to everyone in the second parentheses, and we get $a^3 + a^2b + ab^2$. Next, $-b$ says hello to everyone in the second parentheses, and we get $-ba^2 + (-ab^2) + (-b^3)$. Now, let's add them all together: $a^3 + a^2b + ab^2 + (-ba^2) + (-ab^2) + (-b^3)$. It can be confusing with all the a's and b's, so let's write each term alphabetically; it'll help us recognize like terms. And let's use the doodling technique we used on p. 303, too: $\underline{a^3} + \underline{a^2b} + \underline{ab^2} + (\underline{-a^2b}) + (\underline{-ab^2}) + (\underline{-b^3})$. Wow, check it out: Two sets of terms cancel each other out—the double lines and the wavy lines. Nice! We end up with $a^3 + (-b^3)$, and let's streamline the negative sign for the answer.*

Answer: $a^3 - b^3$

2. $(x - 2)(x + 2 - y)$

3. $(3 + y)(9 - 3y + y^2)$

4. $(a + b)(a^2 - ab + b^2)$

5. $(h^2 + g - 1)(2h^2 - 2g + 3)$ *(Hint: see p. 316.)*

(Answers on p. 410)

. . . .

* This is a special polynomial pattern that you'll learn more about in Algebra II. It's called the "difference of cubes": $(a - b)(a^2 + ab + b^2) = a^3 - b^3$. A similar pattern, called the "sum of cubes," is in problem 3 & problem 4 above.

Special Polynomials

These are very handy to be familiar with, especially when we get into factoring in the next chapter. The last one is my favorite because when we multiply it out, a bunch of stuff cancels away, and that's always so nice, isn't it?

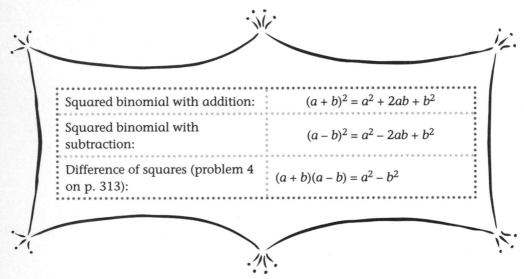

Squared binomial with addition:	$(a + b)^2 = a^2 + 2ab + b^2$
Squared binomial with subtraction:	$(a - b)^2 = a^2 - 2ab + b^2$
Difference of squares (problem 4 on p. 313):	$(a + b)(a - b) = a^2 - b^2$

There's no need to memorize these . . . yet. For now, just keep multiplying everything out and using FOIL. But keep an eye on them, because recognizing these is a lifesaver on tests in Algebra I and Algebra II. Teachers totally love these.

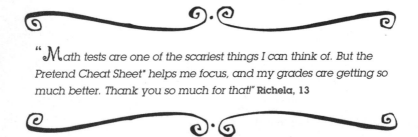

"Math tests are one of the scariest things I can think of. But the Pretend Cheat Sheet* helps me focus, and my grades are getting so much better. Thank you so much for that!" **Richela, 13**

. . . .

* For the Pretend Cheat Sheet and more, check out the Math Test Survival Guide at the end of *Kiss My Math*, and never dread test-taking again!

More than Two Factors

If you ever have to multiply three or more factors, just do them *two at a time*.

$$3x\left(x + \frac{5}{3}\right)(x - 2)$$

Notice there are *three* factors: $3x$, $\left(x + \frac{5}{3}\right)$, and $(x - 2)$. Let's multiply the first two and pretend nothing else exists: $3x\left(x + \frac{5}{3}\right) = 3x^2 + \frac{15x}{3} = \mathbf{3x^2 + 5x}$. Good, now we can rewrite our problem with just two factors:

$$(3x^2 + 5x)(x - 2)$$

We used FOIL to multiply this on p. 311 and saw that it equals $\mathbf{3x^3 - x^2 - 10x}$. Not so bad, right? Just take 'em two at a time.

What's the Deal?

Why does FOIL work? Let's do this problem from p. 313, $(x + 2)(3y^2 - z)$, but this time without FOIL. Instead, we'll only use the distributive property. Just for fun, let's substitute ✿ for $(x + 2)$. Now our expression becomes:

$$✿(3y^2 - z)$$

Distributing the flower, we get ✿ $\cdot 3y^2 - $ ✿ $\cdot z$, which we can rewrite as $\mathbf{3y^2} \cdot$ ✿ $- \mathbf{z} \cdot$ ✿. Substituting $(x + 2)$ back in for the flower, this is:

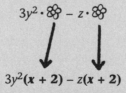

$$3y^2(x + 2) - z(x + 2)$$

Then we just distribute again and simplify:

$$= \mathbf{3xy^2 + 6y^2 - xz - 2z}$$

It's the same answer we got on p. 313 when we used FOIL! See, there's nothing magical going on with FOIL; it's just a compound form of the distributive property in disguise.

Takeaway Tips

 To multiply two binomials together, remember FOIL: **F**irst, **O**utside, **I**nside, **L**ast. Make sure to bring negative signs along for the ride.

 To square a binomial, write it as two binomials side-by-side and then use FOIL.

Use doodling and creative underlines so you can combine like terms more easily.

Remember to leave answers in descending order with streamlined negatives.

TESTIMONIAL

Joslyn Davis
(Downey, CA)
<u>Before</u>: Needed help in math
<u>After</u>: Model, actress, and TV reporter!

        In high school, I wanted to do it all. I was on the cheer team, in student government, dance groups, and more. I hung out with what people probably considered the popular crowd and dated the head of the baseball team. But I was also obsessed with getting A's, and math never came easy to me. In order to "do it all," it seemed I was going to need a little help! And to be completely honest, I was far more interested in the complexities of the kids in my honors and AP classes than in the silly antics of the so-called popular kids. Now I had a great reason to hang out with them. I started asking the smartest kids to study with me—it was almost like getting an inside scoop! Well, my strategy paid off big time because I got to know some truly interesting, fun classmates, and they really did help me understand math, which I will forever be grateful for.

"I wasn't afraid to ask others for help."

Turns out that my curiosity and knack for interviewing and asking the right questions led to a successful career in entertainment reporting! I've worked with ClevverTV, HollywoodTV, ProjectRunway.com, and BlendInTV, just to name a few, and I also model and act in commercials occasionally. Balancing everything while paying the bills is like running my own mini business. Working in entertainment sounds glamorous, but believe me—if you want to survive and thrive, you have to be savvy. Being cute and having an awesome personality will just not cut it.

For example, I'll ask myself how much money I'm investing in my career. Then I have to balance what returns I'll see from those investments, which ultimately helps me make wise decisions. Important choices have to be made, and it's often about the numbers, like one big math problem. Seriously, intelligence—coupled with hard work and tenacity—is what will make or break you in this business.

Even in my personal life, math comes up all the time. After watching Project Runway and doing correspondence work for the site, I decided to take up sewing, which sounds like a completely creative art, but it's so mathematical! Everything from cutting out patterns to trimming thread requires calculations. I've had a few instances of running out of fabric or thread, and I've also had a few disasters where the item came out totally wrong. So now I'm slowing down and making those calculations a little more carefully! I also love to travel to foreign countries, and being able to accurately calculate exchange rates and budget trips is key to not breaking the bank.

Looking back, I'm so glad I challenged myself when I was younger and wasn't afraid to ask others for help. Working up the courage to ask questions isn't always easy, but it's so worth it. Persevering and believing in myself when I came upon a challenge in high school—like a hard math problem—is what has given me the confidence and independence to pursue the crazy career I'm in today!

Chapter 22

The Jewelry Dilemma

Factoring Trinomials with Reverse FOIL and Grouping

*A*h, jewelry. We all have a favorite bracelet or ring—something that works with almost any outfit and just makes us feel happy and sparkly. Depending on the piece, we might actually *be* more sparkly. As time passes on, you will accumulate more bangles, necklaces, earrings, and maybe even toe rings. And with so many options, you might find the task of picking the right jewelry for an outfit to be like solving a puzzle, especially if the top and skirt are pretty different from each other. Wouldn't it be nice if there were a formula to figure out the perfect jewelry every time?

While addressing this important issue, we'll also learn how to factor quadratic trinomials with reverse FOIL. How nice.

Sum & Product Puzzles—and the Perfect Jewelry

Sum & product puzzles look a bit like our girl with the jewelry predicament . . . especially, um, since the girl has numbers on her outfit. But unlike with jewelry,* there's only one right answer, so we'll know when we're done trying options.

Our goal with these puzzles is to find the pair of numbers that *multiply* together to give us the top number (the product) and *add* together to give the bottom number (the sum). Once we figure out what those numbers are, we put them on either side of the X, right where the bracelet and ring would go. Let's do it!

. . . .

* I'll give you some tips on actual jewelry selection on p. 330!

Can we think of a pair of numbers that multiply together to give us 6 and also add together to give us 5? Sure! 2 and 3 will do the job, because $2 \cdot 3 = 6$, and $2 + 3 = 5$. Then we'd just write **2** and **3** on either side of the X.

This was pretty easy, but when the numbers are a little bigger, it's great to have a step-by-step method.

Step By Step

Sum & product puzzles:

Step 1. Look at the <u>top</u> number in the big X. List all the *pairs* of factors that result in this product when multiplied together.

Step 2. Then find the pair that adds together to give you the bottom number.

Step 3. Take the winning pair, and write each number on either side of the big X. Done!

QUICK NOTE   With practice, you'll be able to eyeball most of these puzzles, and you won't need to write out all the factors. But for now and any time you get stuck, write 'em out. Besides, you'll probably get halfway through and find your answer anyway.

Step By Step In Action

Fill in this sum & product puzzle.

Step 1. Okay, what are all the factor pairs of 24? 1 & 24, 2 & 12, 3 & 8, and 4 & 6. We know all of these pairs will multiply to give us 24.

Step 2. Who adds together to give us 11? Aha! Looks like 3 & 8 is the winner.

Step 3. We'll write them on either side of the X. Done!

Answer: $3 \underset{11}{\overset{24}{\times}} 8$

 QUICK NOTE   If the top number (the product) is negative, then *one* of the numbers in the winning pair will be negative. We'll do an example below.

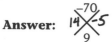 *Take Two: Another Example*

Fill in this sum & product puzzle. $\overset{-70}{\underset{9}{\times}}$

This time, let's try eyeballing it. What is the first pair of factors we think of when we see 70? Well, 7 & 10, naturally! And we know that <u>one of the factors</u> will be negative to give us –70. Hmm, 7 & 10 (one being negative) don't add to give us 9, though. But looking at the prime factorization, $70 = 2 \times 5 \times 7$, we can also find the factor pairs 14 & 5, and 35 & 2. We know these combos each work for the *product*, but what about a sum of 9? Well, $14 + (-5) = 9$, so we've discovered that 14 & –5 is the winning combo!

Answer: $14 \underset{9}{\overset{-70}{\times}} -5$

 QUICK NOTE   If the top number is positive and the bottom number is negative, both "winning" numbers will be negative. For example, with 6 on top and –5 on the bottom, we'd choose –2 & –3, because $(-2)(-3) = 6$ and $-2 + (-3) = -5.$ $-2 \underset{-5}{\overset{6}{\times}} -3$

Do these sum & product puzzles: Find two numbers whose product is the top number and whose sum is the bottom number. I'll do the first one for you.

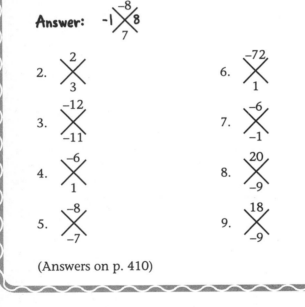

1. $\times$ (top: −8, bottom: 7)

Working out the solution: What are all the factor pairs of 8? Let's see: 1 & 8 and 2 & 4. We know one of the factors needs to be negative, because the product is −8, and they also must add up to 7. How about −1 & 8? Yep: −1 + 8 = 7. Done!

Answer: −1 $\times$ 8 (top: −8, bottom: 7)

2. $\times$ (top: 2, bottom: 3)

3. $\times$ (top: −12, bottom: −11)

4. $\times$ (top: −6, bottom: 1)

5. $\times$ (top: −8, bottom: −7)

6. $\times$ (top: −72, bottom: 1)

7. $\times$ (top: −6, bottom: −1)

8. $\times$ (top: 20, bottom: −9)

9. $\times$ (top: 18, bottom: −9)

(Answers on p. 410)

Factoring with Reverse FOIL

In Chapter 21, we learned the FOIL method of multiplying two binomials. We just think to ourselves, "First, Outside, Inside, Last" and we can easily multiply things like this:

$$(3x - 4)(x + 3) \;=\; 3x^2 + 9x - 4x - 12 \;=\; \mathbf{3x^2 + 5x - 12}$$

Now we're going to go in reverse. Yes, just like with reverse distribution, we spent all that time learning how to do something and now we're going to learn how to *undo* it.*

We'll get things like $3x^2 + 5x - 12$, and we'll figure out how to write it as a product of factors, so it looks like this: $(3x - 4)(x + 3)$. Now you might be saying, "How could we possibly know *how* to undo the FOIL and figure out how to write it as two sets of parentheses multiplied together?"

By the end of this chapter, missy, you'll be amazed at what you can do. Let's try this one: $\mathbf{x^2 + 6x + 8}$.

Ever seen those websites where they show you what celebrities looked like as a baby, sometimes when they were still in *diapers*? Well, we're going to go back in time and figure out what $\mathbf{x^2 + 6x + 8}$ looked like when it was still in diapers, I mean, in *parentheses*. I bet she'll be so cute.

In the first position of both sets of parentheses, we should use x, because in FOIL, when the **F**irst multiplication happened, we know it resulted in x^2. Makes sense, right?

$$\mathbf{(x + \,?\,)(x + \,?\,)} \;=\; x^2 + 6x + 8$$

How do we fill in the mystery numbers in the last two slots? Thinking again about what must have happened during FOIL, in the **L**ast step, those two mystery numbers multiplied to be **8**. And as it turns out, we need this pair of numbers to not only *multiply* to be 8, but also *add* up to be **6** (the coefficient on the middle term). Sound familiar? Yep—it's just like a sum & product puzzle! Finding the pair of numbers whose product is 8 and whose sum is 6 is easy; it's **2** & **4**. So let's stick 'em in:

$$(x + \mathbf{2})(x + \mathbf{4}) \;=\; x^2 + 6x + 8$$

Multiplying out the left side with FOIL to check our answer, we get $x^2 + 4x + 2x + 8 = \mathbf{x^2 + 6x + 8}$. Nice! We've gone back in time, and yep, she looked pretty adorable when she was still in parentheses.

Answer: $(x + 2)(x + 4)$

. . . .

* C'mon, there are lots of things we do, only to undo them later. You know, making the bed, dressing & undressing (helpful for shower time) . . .

QUICK NOTE   If a mystery number turns out to be negative, then there will be subtraction inside its parentheses instead of addition. So moving forward, we won't write in the + (or −) until we're writing in the mystery numbers, too.

Step By Step

Factoring with reverse FOIL when x^2's coefficient is 1:

Step 1. Make sure you have an expression in the form $x^2 + bx + c$, where b and c are integers. (If b or c has subtraction before it, you should consider that to be a negative value for the number.)

Step 2. Draw a set of parentheses like this, with blank spots where the mystery numbers and signs will go: $(x\quad)(x\quad)$.

Step 3. We need to find the pair of numbers whose product is c (the constant) and whose sum is b (the coefficient of the middle term). Start by listing all of c's factor pairs, just like in the sum & product puzzles, and find the winning mystery pair.

Step 4. Fill in the mystery pair in the blank spots, and if a mystery number is negative, write it as <u>subtraction</u> inside its parentheses.

Step 5. Check your answer by using FOIL to multiply it out. Done!

And... Action! Step By Step In Action

Factor this expression: $x^2 - x - 6$.

Step 1. It's in the correct form, and in this case, $b = -1$ and $c = -6$.

Step 2. Let's draw our parentheses: $(x\quad)(x\quad)$

Step 3. The two mystery numbers must multiply to give us −6 and add to give us −1. Hmm, what are −6's factor pairs? They are −6 & 1, 6 & −1,

3 & –2, and –3 & 2. Do any of these pairs add up to give us –1? Yes, **–3 & 2**, because –3 + 2 = –1. Yay!

Step 4. We'll write them in and convert –3's negative sign into subtraction in its parentheses: $(x - 3)(x + 2)$

Step 5. Using FOIL to multiply this out, let's make sure we get what we started with:

$$(x - 3)(x + 2) \ = \ x^2 + 2x - 3x - 6 \ = \ \textbf{x}^2 - \textbf{x} - \textbf{6}$$

Yep, we got it right!

Answer: $(x - 3)(x + 2)$

QUICK NOTE   We can always rewrite subtraction as "adding a negative" to more easily see b's and c's values. In the last example, we could have first written $x^2 - x - 6$ as $x^2 + (-x) + (-6)$, or even included the sneaky invisible coefficient, too: $x^2 + (-1x) + (-6)$.

Doing the Math

Factor these polynomial expressions using reverse FOIL. I'll do the first one for you.

1. $y^2 + 9y - 70$

<u>Working out the solution</u>: Let's write our parentheses: $(y \quad)(y \quad)$. So, we need to figure out a pair of numbers whose product is –70 and whose sum is 9, right? We actually did this sum & product puzzle on p. 325, and we found that pair of numbers is **–5** and **14**, since $(-5)(14) = -70$ and $-5 + 14 = 9$. Let's fill 'em in: $(y - 5)(y + 14)$. Multiplying it out, we get $(y - 5)(y + 14) = y^2 + 14y - 5y - 70$. Combining

the middle terms, we get exactly what we started with: $y^2 + 9y - 70$. So we did the reverse FOIL right!

Answer: $(y - 5)(y + 14)$

2. $x^2 + 8x + 7$

3. $w^2 + w - 12$

4. $h^2 - 12h + 11$

5. $x^2 + 2x - 15$

6. $y^2 - 2y - 15$

7. $x^2 - 11x + 18$

8. $g^2 + 7g - 18$

9. $x^2 - 9x + 18$

(Answers on p. 410)

The Whole Point of This

Getting good at factoring will enable us to solve a whole bunch of quadratic *equations*, which can be very helpful in the real world for designing and building things. Be sure to check out the Reality Math on p. 361.

Danica's Diary

Here's my advice on the jewelry: Classic, tasteful stuff always works best. And how to choose? Start paying attention to what you like on other people and *experiment*. That's what I do. Personally, I love gold or silver hoop earrings, as long as I'm not wearing a big flashy necklace. If so, then I opt for small studs or drop earrings. And the best accessory of all? The *confidence* that comes from feeling smart and capable. And the more you challenge yourself, like by tackling hard math, the more confidence you'll have. Gorgeous!

When the x^2 Coefficient Isn't 1

When the coefficient of x^2 isn't 1, like with $3x^2 - 5x - 2$, things become a bit more complicated. See, we can't start out our factoring like this $(x \quad)(x \quad)$, because we'd be ignoring the 3. In this case, it should actually start out like this $(3x \quad)(x \quad)$, so that the first two terms multiply to become $3x^2$.

Time to find our two mystery numbers and fill them in. As always, their product must be –2. However, they will NOT add up to –5. This is because the 3 interferes and messes it up. So, how do we proceed? I like to call this "plug and play." We plug in different options and play around until something works. Very scientific, I know.

$$(3x \quad)(x \quad) = 3x^2 - 5x - 2$$

Hmm, we know the mystery numbers' product is –2, so we could use –2 & 1 or 2 & –1. It might seem like there would be only two options to check, but there are *four* options, because we don't know if the 2 gets paired with the $3x$ or the x. Here are the four options:

$$(3x + 2)(x - 1) \quad (3x - 2)(x + 1) \quad (3x - 1)(x + 2) \quad (3x + 1)(x - 2)$$

Let's multiply out the first option and see what we get:

$$(3x + 2)(x - 1) = 3x^2 - 3x + 2x - 2 = 3x^2 - x - 2$$

Hmm, wrong middle term. After multiplying them all out, we find that the last option is the only one that will give us the correct middle term, –5x:

$$(3x + 1)(x - 2) = 3x^2 - 6x + x - 2 = 3x^2 - 5x - 2$$

Answer: $(3x + 1)(x - 2)$

Hey, if it were easy, everybody'd be doing it. Be strong!

Factoring with reverse FOIL when x^2's coefficient is _not_ 1: Plug and play

Step 1. Make sure you have an expression in the form $ax^2 + bx + c$, where a, b, and c are nonzero integers with no common factors. If they have a common factor, then factor it out. If a is negative, factor out –1 from the whole thing.

Step 2. Draw two sets of parentheses with blank spots where the mystery numbers go: (__x)(__x). If you factored out –1, then your parentheses should look like this: (–1)(__x)(__x). If you factored something else out, it should be stuck to the outside, too!

Step 3. For the blank spots *before* each x, list the possible options: What are the factor pairs of a? Write down those options; if there's only one factor pair, go ahead and write it inside the parentheses! (Remember, a coefficient of 1 usually isn't written, so if your factor pair for a is 5 & 1, you'd have $5x$ and x in your first two positions.)

Step 4. For the second blank spots in the parentheses, we know the mystery numbers will multiply to be c (which is negative if it has subtraction before it). List all options—what are the factor pairs for c?

Step 5. Write down the possible combinations, and find the one that multiplies out to give the original expression.

Step 6. Check your work by using FOIL if you haven't already. Done!

Just follow along; you'll get the hang of it soon.

Let's factor $2x^2 - 13x - 7$.

Steps 1–3. It's in the right form, so let's write our parentheses. Since the only factor pair for 2 is 2 & 1,* we know we'll need $2x$ and x in the first spots: $(2x\ \ \ )(x\ \ \ )$.

Step 4. The only possible options for the second slots are –7 & 1 and 7 & –1.

Step 5. Let's write out all possible combinations, looking for the middle term, **–13x**.

$(2x - 1)(x + 7)\ =\ 2x^2 + 14x - x - 7\ =\ 2x^2 + 13x - 7$ Nope.

$\mathbf{(2x + 1)(x - 7)}\ =\ 2x^2 - 14x + x - 7\ =\ \mathbf{2x^2 - 13x - 7}$ Yes!

$(2x + 7)(x - 1)\ =\ 2x^2 - 2x + 7x - 7\ =\ 2x^2 + 5x - 7$ Nope.

$(2x - 7)(x + 1)\ =\ $ Hey wait, no need to do this extra work. We're done!

Answer: $(2x + 1)(x - 7)$

"$\mathcal{S}$ometimes I 'get' things out of order. Like, I don't understand the beginning of a math section, but then something at the end of it makes the whole thing click." **Tabitha, 17**

Next, I'll show you two strategies I like to use to make things go faster. . . .

. . . .

* Advanced footnote: We don't need to consider the option of –2 & –1 for the spots <u>before</u> the *x*, because in the final answer, we could always just factor out –1 from both parentheses (which would cancel each other), and simply change the sign of the second slots. So, we let the "signage" get taken care of in the *second* slots.

Shortcut Alert: Two Strategies

Strategy 1: Temporarily ignore the signs!

Instead of writing out each and every single option, we can start off by leaving blank spots where the + or − should go and just write down the options for the numbers and variables. This cuts the writing in half! In the above example, $2x^2 - 13x - 7$, knowing that the parentheses will have $2x$ and x in their first slots, and looking at the 7, we could think, Okay, what are 7's factors? Just 7 & 1. So we would write:

$$(2x \quad 1)(x \quad 7) \qquad (2x \quad 7)(x \quad 1)$$

Then we can get an overview of <u>all</u> the options without so much writing . . . and we're ready for the second strategy.

Strategy 2: Get hints from the middle term and reason it out!

Looking at our handwritten options above and knowing that eventually those parentheses will multiply out to give <u>the middle term −13x</u>, we might notice that it could be awfully good if the $2x$ and -7 ended up multiplying times each other . . . because that will give $-14x$, which is close to $-13x$, and there's only one scenario that does this:

$$(2x + 1)(x - 7)$$

And that's our answer! This kind of logic can save you a ton of time, and as a bonus, you'll sharpen those logical problem-solving skills that will make you one clever kitty.

I'll do the next example with these shortcuts. . . .

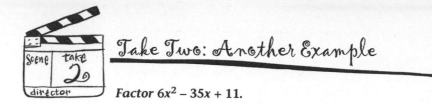

Take Two: Another Example

Factor $6x^2 - 35x + 11$.

Steps 1 and 2. Although the numbers are big, there's no GCF to pull out from every term. Time to write the parentheses:

$$(\quad x \quad)(\quad x \quad)$$

Step 3. What are the factor pairs for the first slots? This time, x^2's coefficient is 6. It isn't prime, so we have some options: We could use $2x$ & $3x$ or $6x$ & x. Drat, okay fine.

$$(2x \quad)(3x \quad) \qquad (6x \quad)(x \quad)$$

Step 4. Forgetting about the signs for a moment, what are the factor pair options for the second slots? Luckily, 11 is prime, so the only options are 1 & 11:

$$(2x \quad 1)(3x \quad 11) \qquad (2x \quad 11)(3x \quad 1)$$
$$(6x \quad 1)(x \quad 11) \qquad (6x \quad 11)(x \quad 1)$$

How about the signs? Since the last term of the original polynomial is positive, we know that <u>both parentheses will have + or both parentheses will have –</u>. And we end up with eight options to try. Yikes!

Step 5. Before we freak out, let's notice that we need to somehow get that middle term, $-35x$. <u>So the parentheses will both have – not +.</u> Imagining doing FOIL with our above options: If the $6x$ and 11 multiplied times each other, we'd get $66x$, which is huge. But if the $3x$ and the 11 multiplied times each other, we'd get $33x$, which is close to $35x$. So let's write parentheses where the $3x$ and 11 multiply times each other: $(2x \quad 11)(3x \quad 1)$. But since $-35x$ is negative, we'll try the option that uses negative signs:

$$(2x - 11)(3x - 1)$$
$$= 6x^2 - 2x - 33x + 11$$
$$= 6x^2 - 35x + 11$$

Yes, we found it!

If this hadn't worked, then we'd just go back and try some other options. But we sure made an educated guess about which to try first and saved a whole bunch of time.

Answer: $(2x - 11)(3x - 1)$

Doing the Math

Factor these using reverse FOIL. I'll do the first one for you.

1. $-12x^2 + 70x - 22$

<u>Working out the solution</u>: Hmm, this is a bit of a mess. Let's factor out -2 and make life simpler. Using reverse distribution, we get $-12x^2 + 70x - 22 = -2(6x^2 - 35x + 11)$. On p. 335, we found that $6x^2 - 35x + 11 = (2x - 11)(3x - 1)$, so for the full answer, we just stick the -2 back on.

Answer: $-2(2x - 11)(3x - 1)$

2. $2x^2 + 3x + 1$

3. $7h^2 + 15h + 2$

4. $6w^2 + 5w + 1$

5. $2x^2 - 3x - 2$

6. $10m^2 + 33m + 9$

7. $-10y^2 + 15y + 10$

8. $4g^2 - 3g - 1$

9. $3x^2 + 8x - 3$

(Answers on pp. 410–11)

What's the Deal?

There's an inverse relationship between multiplying and factoring that I want to point out. The following chart will help you get a wider perspective of what we're doing.

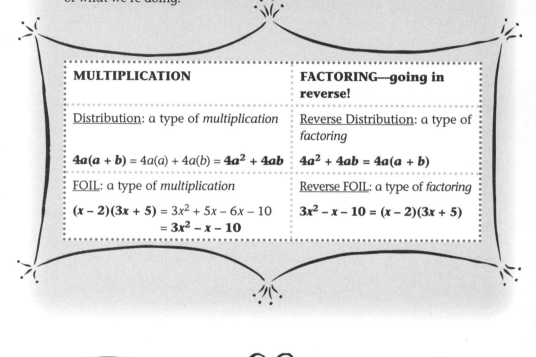

MULTIPLICATION	FACTORING—going in reverse!
<u>Distribution</u>: a type of *multiplication*	<u>Reverse Distribution</u>: a type of *factoring*
$4a(a + b) = 4a(a) + 4a(b) = \mathbf{4a^2 + 4ab}$	$\mathbf{4a^2 + 4ab = 4a(a + b)}$
<u>FOIL</u>: a type of *multiplication*	<u>Reverse FOIL</u>: a type of *factoring*
$\mathbf{(x - 2)(3x + 5)} = 3x^2 + 5x - 6x - 10$ $= \mathbf{3x^2 - x - 10}$	$\mathbf{3x^2 - x - 10 = (x - 2)(3x + 5)}$

"*What I like about math is that there is only one right answer, and I get really proud of myself when I figure out what that answer is.*" **Jacqueline, 14**

Factoring by Grouping

Ah, factoring by grouping. It's a bit more advanced, but you can totally handle it. And . . . it can be a *lifesaver* for when there are a ton of factor options, like $12x^2 - 17x + 6$. Um, with that 12 and the 6, there are gazillions of options (12 options, or up to 48 options if we didn't notice that both parentheses must use subtraction—yikes!). Can you imagine trying to factor this with the techniques we've been using?

Here's the idea behind grouping: See, once upon a time, $12x^2 - 17x + 6$ was two parentheses multiplied together, right? And after FOIL, there were *two* middle terms. Then someone combined those middle terms, and their sum was $-17x$. We'll figure out what those two middle terms were *before* they were combined! Next, we'll flex our GCF factoring skills a couple of times, and voilà, done! Not so bad, right? As an added bonus, when you do Step 5 below, you'll actually feel your brain expand.*

Step By Step

Factoring quadratic trinomials by grouping:

Step 1. We'll need the expression to be in the form $ax^2 + bx + c$, where a, b, and c are nonzero integers with no common factors and a is <u>positive</u>. If they have a common factor, then factor it out. If x^2's coefficient is negative, factor out -1 from the whole thing. This step is crucial! Now that a is indeed positive, it's time to assign a, b, and c's values.

Step 2. Now, we want a new pair of numbers whose product is $\boldsymbol{ac}$ and whose sum is $\boldsymbol{b}$.

Step 3. Take those two new numbers and <u>stick the variable on each of them</u>. Rewrite the original expression, <u>splitting the middle term into these two new terms</u>.

Step 4. Find the GCF of the first two terms only and factor it out. Do the same with the second set of terms. If the <u>second</u> set of terms <u>starts</u> with subtraction, then be sure the GCF you pull out is negative, too. At this stage, we have only two big "terms" in our expression, each stuck to an identical parentheses. (If they're not identical, try factoring out -1 from one.)[†]

Step 5. Now the identical sets of parentheses *themselves* are a common "factor" for our two big terms! So <u>factor it out</u>, and now there should just be two sets of parentheses multiplied together.

Step 6. Check your work by multiplying them out. Done!

This will make much more sense in an example. Let's do one!

. . . .

* In a good way.
[†] See problem 1 on p. 341 for an example of having to factor out -1 like this.

Factor $12x^2 - 17x + 6$.

Step 1. Looks good: x^2's coefficient is positive, and there's no GCF.

Step 2. What's ac? In this case, $a = 12$ and $c = 6$, so $ac = 12 \cdot 6 = $ **72**. And $b = $ **–17**, so we'll look for the pair of numbers whose product is 72 and sum is –17. What are 72's possible factor pairs? Well, the first pair that comes to mind is from our times tables, 8 & 9. And hey, if they're both negative, their sum is –17! So we get **–8** and **–9**.

Step 3. Sticking x to them, we'll use $-8x$ and $-9x$ to rewrite our expression with the new middle term split up: $12x^2 - 8x - 9x + 6$. This is exactly equal to our original expression.

Step 4. Looks like the GCF of the <u>first two</u> terms is $4x$. Factoring that out, our expression becomes:

$$12x^2 - 8x - 9x + 6 \quad = \quad \mathbf{4x}(3x - 2) - 9x + 6$$

The GCF of the last two terms is 3, but since there's subtraction before the $9x$, we should factor out the negative **–3**; otherwise the negative signs will be too hard to keep track of:

$$4x(3x - 2) - 9x + 6 \quad = \quad 4x(3x - 2) - \mathbf{3}(3x - 2)$$

Notice how when we factored out –3, the 2 had to become negative. This step can be tricky, so read it over and make sure you understand it, especially the negative signs!

Step 5. Notice that our expression now has a common factor: $(3x - 2)$. Strange, huh! And when we "factor" it out with reverse distribution, our leftover terms are $4x$ and –3. Check it out:

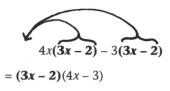

$$4x(\mathbf{3x - 2}) - 3(\mathbf{3x - 2})$$

$$= (\mathbf{3x - 2})(4x - 3)$$

It looks like we've ended up with our answer.*

· · · · · · · · · ·

* Was I right? Doesn't this step expand your brain a little?

Step 6. To check our work, we'll multiply it out with FOIL and hope to get what we started with.

$$(3x - 2)(4x - 3) \;=\; 12x^2 - 9x - 8x + 6 \;=\; \mathbf{12x^2 - 17x + 6}$$

Yep! We got it right.

Answer: $(3x - 2)(4x - 3)$

Now, could we have found that answer by just writing out options and testing? Sure . . . but that could have taken forever with a problem like this, and now you've got another tool under your belt—grouping!

QUICK NOTE   It doesn't matter in which order you write the split-up middle terms. It will work out either way. If we'd written it with the −9x before the −8x, like this, $12x^2 - 9x - 8x + 6$, then the GCF of the first two terms is 3x, and the GCF of the second two terms is 2. We'd have gotten $3x(4x - 3) - 2(4x - 3)$. Then we would have factored out the $(4x - 3)$ and gotten the same answer, but written in the opposite order: **$(4x - 3)(3x - 2)$.** Ta-da!

When I was in school, I remember thinking that grouping seemed like magic—how the parentheses end up being identical and everything. So surely, it could only work on very special problems. But I was wrong! Really, try this method with any of the factoring problems from this chapter and it will work. You just might have to use "1" or "−1" as a GCF in Step 4. We'll do this in problem 1 on p. 341.

QUICK NOTE   When factoring out the GCF from the second set of terms in Step 4, if the second set starts with a negative sign and you forget to factor out the negative, you can still do it later. I'll show you how this works below in problem 1.

Doing the Math

Factor these problems using the grouping method. I'll do the first one for you.

1. $2x^2 - 3x + 1$

<u>Working out the solution</u>: Here's an example of the grouping technique on an "easy" problem, which can actually be a little trickier. So $ac = 2$ and $b = -3$, and the pair whose product is 2 and sum is -3 would be **−2 & −1**. Rewriting the middle term, $-3x$, as $-2x - 1x$, our expression becomes $2x^2 - 2x - 1x + 1$. Pulling out the GCF from the first two terms, which is $2x$, and the GCF from the last two terms, which is 1, we get **$2x(x - 1) + 1(-x + 1)$**. I know that was weird to use 1 as a GCF, but it's legal, after all. Hmm, our parentheses aren't identical. If we factor out **−1** from the second parentheses,[*] it will look like $(x - 1)$ instead of $(-x + 1)$, and we'll end up with subtraction *outside* the parentheses instead of addition: **$2x(x - 1) + 1(-x + 1) = \underline{2x(x - 1) - 1(x - 1)}$**. And now the GCF of these two terms is $(x - 1)$. Let's pull that out: $(x - 1)(2x - 1)$. And yes, if we multiplied this out, we'd get $2x^2 - 3x + 1$. Voilà!

Answer: $(x - 1)(2x - 1)$

2. $12x^2 - x - 6$

3. $10x^2 + 11x - 6$ *(Hint: How many ways can you factor 60?)*

4. $3x^2 - 5x - 2$ *(Hint: Split −5x into −6x + 1x and use the 1 coefficient like I did in problem 1.)*

5. $8x^2 + 2x - 15$

(Answers on p. 411)

• • • • • • • • • •

[*] Factoring out −1 means that each term's sign changes.

For more practice, do the problem set from p. 330, using *grouping* as the technique—you should get the same answers!

Takeaway Tips

Reverse FOIL can be used to *factor* quadratic trinomials; in other words, to rewrite them as two sets of parentheses multiplied together.

When x^2's coefficient is 1, we need to find a pair of mystery numbers whose *product* is the constant and whose *sum* is the middle term's coefficient.

When x^2's coefficient is not 1, then things are a bit harder. But the middle term can still give us clues about which numbers to try first.

Grouping is all about rewriting the middle x term as the sum or difference of two terms and then finding a bunch of GCFs.

Always check your work: Just multiply and make sure you got what you started with.

Chapter 23

Boyfriend or "Special" Friend

Factoring Special Polynomials

$\mathscr{S}$omeday when you get a boyfriend—and believe me, there's no rush—it will be very interesting to see how your parents react. And when I say interesting . . . well, let's just say I have a little story on p. 346 in case you want to read about my "special" friend.

Factoring Special Polynomials

In some cases, we can avoid doing the reverse FOIL factoring from the last chapter. Remember the chart from p. 318? Well, here it is again, in reverse order! And we can use this chart to recognize these patterns in certain problems in homework and tests, which makes 'em a breeze. These are called "special polynomials." They really are pretty special.

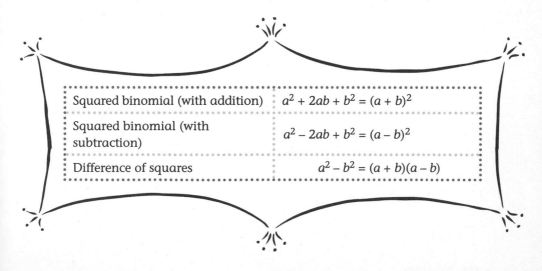

Squared binomial (with addition)	$a^2 + 2ab + b^2 = (a + b)^2$
Squared binomial (with subtraction)	$a^2 - 2ab + b^2 = (a - b)^2$
Difference of squares	$a^2 - b^2 = (a + b)(a - b)$

Teachers just love the *difference of squares* one, so keep an eye out for it!

You'll come across many problems that fit into one of these patterns, for example, $x^2 - 36$, which takes the form $a^2 - b^2$. It's so nice when it happens. It's like finding a friend in the middle of a crowded room of strangers.*

Looking at the "difference of squares" in the chart and substituting x for a and 6 for b, we see that because $a^2 - b^2 = (a + b)(a - b)$, it should be true that $x^2 - 36 = \boldsymbol{(x + 6)(x - 6)}$. But does this work? Let's try it: $(x + 6)(x - 6) = x^2 - 6x + 6x - 36 = \boldsymbol{x^2 - 36}$. Yep, it does!

And these patterns are a lifesaver for monster problems like $\boldsymbol{25x^2 - 30x + 9}$. I mean, look at those numbers! Looking at the chart, here's how your thinking might go: "Hmm, $25x^2$ is a perfect square, and so is 9. I wonder if this takes the form of the special polynomial $a^2 - 2ab + b^2$?

Let's see: $25x^2$ would have to be the a^2, which means $\boldsymbol{a \to 5x}$, right? Also, since $b^2 \to 9$, $\boldsymbol{b \to 3}$. Then filling those values into $2ab$, we get: $2ab \to 2(5x)(3) = \boldsymbol{30x}$. Yes, it's a match!" Check it out:

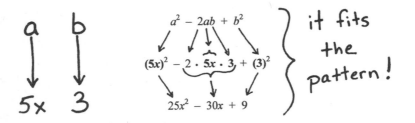

Since it fits the pattern with $a = 5x$ and $b = 3$:

$$a^2 - 2ab + b^2 = (a - b)^2$$

$$\boldsymbol{25x^2 - 30x + 9 = (5x - 3)^2}$$

And it's easy to check our work by multiplying it out with FOIL:

$$\boldsymbol{(5x - 3)^2} = (5x - 3)(5x - 3) = 25x^2 - 15x - 15x + 9 = \boldsymbol{25x^2 - 30x + 9}$$

Yep, we got it right!

Answer: $(5x - 3)^2$

---

* Don't believe me? Just wait till you spot one on a test for the first time. You'll see.

QUICK NOTE   Don't worry if that moved too quickly. With practice, you'll be a champ at recognizing these patterns. For now, use the chart on p. 343 as you do the problems below. For each one, just try to figure out what *a* and *b* have to be in order for it to match one of the patterns. Some might be hidden because of a GCF that needs to be pulled out.

Doing the Math

Factor the following expressions, using the chart on p. 343. I'll do the first one for you.

1. $32gh^3 - 98gh$

<u>Working out the solution</u>: Hmm, looks a bit strange, but let's just continue as normal: Is there a GCF to be pulled out with reverse distribution? Yes! We'll pull out $2gh$ from each term, so it becomes $2gh(16h^2 - 49)$. Aha! Inside the parentheses sure looks like a difference of squares, $a^2 - b^2$, where $a \rightarrow 4h$ and $b \rightarrow 7$. According to the chart, that means $16h^2 - 49 = (4h + 7)(4h - 7)$, so the entire thing looks like $2gh(16h^2 - 49) = \mathbf{2gh(4h + 7)(4h - 7)}$. Let's check our work by multiplying it out: $2gh(4h + 7)(4h - 7) = 2gh(16h^2 - 28h + 28h - 49) = 2gh(16h^2 - 49) = \mathbf{32gh^3 - 98gh}$. Yep!

Answer: $2gh(4h + 7)(4h - 7)$

2. $x^2 - 9$ 6. $9x^2 - 12x + 4$

3. $2x^2 - 18$ 7. $5xy^2 - 45x$

4. $4w^2 - 25$ 8. $16h^2 + 8h + 1$

5. $12w^2 - 75$ 9. $27x^3 - 36x^2 + 12x$

(Answers on p. 411)

Danica's Diary
BOYFRIEND VS. SPECIAL FRIEND

When I was 17, I had my first long-term
boyfriend, Matt. I'll never forget the first
time my dad introduced him to someone else.
He said, "And this is Danica's...
(uncomfortable pause—then, slowly)...
special friend." I think he just couldn't
bring himself to say "boyfriend," even though we'd
been dating exclusively for like 4 months at that
point.

I'm not sure I can adequately portray just how
mortified I was, as Matt attempted to lamely shake
hands with whoever it was. How could he hold his
head up high, having been given that dubious title
of "special friend"? Just picture that for a second.
Lame, lame, lame. Of course, I was way too embarrassed
and shy to correct my dad. It was awful. I know my
dad didn't mean any harm, but if he really couldn't
admit that his baby girl was dating, he could have
just said "friend," and it would have been so much
better!

Matt and I ended up laughing about the whole thing
later. It actually became an inside joke between us,
and he didn't hold a grudge at all. That's part of how
I knew Matt was a good guy. It just goes to show that
no matter how much your parents might embarrass you in
front of the guy you like, there's a silver lining:
It's a good opportunity to judge his character.

Memorizing the Patterns

Just imagine how daunting most of those factoring problems on p. 345 would have seemed if we hadn't had our chart to look at! To help yourself memorize the patterns, notice that their factored forms (the right sides, below) all look pretty similar, if you don't use exponent notation:*

$$a^2 + 2ab + b^2 = (a+b)(a+b)$$

$$a^2 - 2ab + b^2 = (a-b)(a-b)$$

$$a^2 - b^2 = (a+b)(a-b)$$

In fact, the only difference is the *signs* inside the parentheses. We've got plus/plus, then minus/minus, then plus/minus. A great way to practice learning the left sides is to randomly write out one of the *right* sides of the above identities, and multiply it out. You can do it any time you have a minute or two to spare and a piece of paper and a pencil!

The next time your teacher says, "Excuse me, miss, but are you doodling again?" You can say, "Deepest apologies, professor. I was just deriving the difference of squares." (British accent optional.)

Takeaway Tips

Memorizing the special polynomials is more helpful than you'd think. Learn them by multiplying out $(a + b)(a + b)$, $(a - b)(a - b)$, and $(a + b)(a - b)$.

Always start factoring by pulling out a GCF if it exists. And doing so might even reveal a hidden pattern!

. . . .

* Remember, $(a + b)^2 = (a + b)(a + b)$ and $(a - b)^2 = (a - b)(a - b)$. When in doubt, multiply it out!

QUIZ:
Are You a Perfectionist?

𝔇o you expect yourself to be perfect, or do you forgive yourself for being, you know, a <u>human being</u>?

Take this quiz by expert psychologist Dr. Robyn Landow and contributor Anne Lowney, and see how you fare!

1. If someone picked up your math notebook and flipped through a few pages, what would she find?

    a. Handwriting of typewriter quality, organized neatly into an outline. There are no mistakes because whenever you mess up, you *have* to rewrite the entire page.

    b. A reasonably neat set of notes with the average amount of revisions, scratch outs, and doodles in the margins.

    c. Papers stuffed throughout—including history homework; old, half-finished assignments; and a collection of your best teacher caricatures. Hey, is that a banana peel?

2. You put your heart and soul into your last science project. When you get a B+, you:

    a. Are pleasantly surprised. Getting a B+ was some sort of miracle. Putting your "heart and soul" into it just meant that you weren't actually texting your friends while you were writing the science report.

    b. Can't help but feel disappointed that you didn't get an A, but it doesn't take long for you to move on; a B+ is a good grade, and you're proud of the work you did.

    c. Feel robbed! That project was flawless, and you deserve nothing but the highest praise for it. You contemplate bringing it up with the teacher—maybe she'll reconsider?

3. At the beginning of each term, you set goals for yourself. What do they look like?

    a. Straight A's, ballet, student council, and softball. (With enough work, you'll make team captain, too!) You've always set the bar high, and everyone thinks of you as Superwoman. You don't want to let everyone down, or yourself.

b. Nothing special. You're coasting on a smooth wave of B's and C's with no particular plan for change. Besides, there's a cute guy you want to be available for, in case he wants to ask you out, like, every night of the week.

c. A set of realistic goals. Maybe to bring your history grade up or go out for the track team. You know that if you plan too many activities, you won't be able to do all of them well. Anyway, nobody is Superwoman, except, well, Superwoman.

4. You are writing a report for English class. It's finished, but you still need to go back and proofread. You:

a. Carefully review your report, noting spelling and punctuation mistakes in your signature purple ink. You revise some sentences to make them clearer and retype it. You've worked hard, and you're happy with the end result.

b. Let the computer do the spell check and hit print. Simple as that!

c. Spend more time editing your paper than you did writing it. The more you revise, the more problems you see: "Argh! A five-year-old could have done a better job!" You're so frustrated that you can't even tell if you're making it better or worse.

5. While taking personality quizzes like this one, you mostly:

a. Try to figure out which answers will make you the "best" quiz taker, score-wise—even though no one else will ever know what your score is.

b. Aren't overly concerned about how you'll do; it's just not that important.

c. Are as accurate as you can be, even if it doesn't paint a pretty picture. After all, you wouldn't be fooling anyone but yourself.

6. You're in an advanced math class this year, and most of the other students are serious about their grades. You:

a. Are conscious of how the other kids in your class are doing and use it to check your own progress. Still, you would never get down on yourself for getting a lower grade than a classmate. Everybody has good days and bad.

b. Constantly measure your success against theirs. No matter how well you do, if Suzie Q got a 95 on the last test, your 94 might as well be an F. You're extremely competitive, and second best just doesn't cut it!

c. Care only about your own grade. Why worry how the other kids are doing? It's not like they're obsessing about *your* scores.

7. Finally, your English teacher assigns a fun project. In groups, you create a skit based on *Treasure Island*, but you were not assigned as group leader, and your group isn't doing as good a job as you'd hoped.

    a. You have a great time working with your classmates and giggling at how bad the swords and costumes look. It would be too much trouble to fix things. So what if it's not professional play quality? This is a blast! A good grade is just a bonus.

    b. You're trying to be patient, but you decide to take charge. After all, those tin-foil swords look totally fake, and why can't the other kids memorize a few simple lines? The group leader seems annoyed with you, but your grade depends on this, and they all need to learn how to do things right.

    c. You want your group to do the best they can, so you offer as many helpful ideas as possible without stepping on the toes of the group leader. Plus, your tiny slipups make the class laugh!

8. You decided to really challenge yourself this term; you're taking all honors classes. But for once, you didn't make honor roll. It's posted for all to see in the school newsletter, and your friends notice it right away. You think:

    a. "Whatever. I don't care what anyone thinks. I don't even care about making honor roll. It doesn't mean anything."

    b. "Maybe they'll also notice that I'm taking all the honors classes, and I'd rather challenge myself and get B's than take easy classes and get A's."

    c. "I feel like a failure. How am I going to explain this to everyone? They know I *always* get good grades. Now what will they think of me? Maybe I should drop out of some hard classes so my grades will be higher."

9. You entered a school-wide short story contest. The votes are in, and you're awarded runner-up—second place. As you're staring at the results, you think to yourself:

    a. "I should have won, and if I had worked even harder, I would have."

    b. "I don't like second place as much as first place . . . but it's definitely cool to get a prize."

    c. "Dang, I'm good. Maybe I'll write a novel next time."

Scoring

1. a = 3; b = 2, c = 1	4. a = 2; b = 1; c = 3	7. a = 1; b = 3; c = 2
2. a = 1; b = 2; c = 3	5. a = 3; b = 1; c = 2	8. a = 1; b = 2; c = 3
3. a = 3; b = 1; c = 2	6. a = 2; b = 3; c = 1	9. a = 3; b = 2; c = 1

If you scored 9–15 . . .

Nope! You're certainly *not* a perfectionist. It's great to be a laid-back person, but be careful that you aren't selling yourself short by not setting goals to achieve. Again, it's fabulous that you are so relaxed, but maybe it's time to set some bigger goals and go after them. What are you good at? What motivates you? What do you want out of your life, both personally and at school?

Once you've come up with a list, ask yourself what steps you need to take to reach your goals and what you need to do to achieve them. There's an important distinction between "calm" and "lazy," after all!

More importantly, ask yourself if there are any roadblocks in your life that might be preventing you from setting or reaching these goals. Are you hanging around with friends who are working hard and who are proud of their achievements? If you're not bringing people into your life who have similar values and priorities, it'll be an uphill battle for you.

You'll benefit from striving a little more in your life. You just might be surprised at how good it feels and what you might accomplish along the way!

If you scored 16–21 . . .

You've got a good balance. You might get down on yourself sometimes, but your whole life is not about being perfect. Of course, you love to perform at the top of your class and get A's, but your balanced attitude means you don't freak out when things don't come out entirely as you want them to!

As life gets more complicated and your goals increase, you can keep this healthy attitude by making sure to always reward yourself for what you do accomplish. Keep enjoying the journey without getting overly hung up on the results.

With your healthy attitude, there's no stopping you! Go ahead and set a goal that really motivates you. Of all people, *you* know it's okay if you don't reach that goal exactly as you wanted. "Shoot for the moon— even if you miss, you'll land among the stars."

If you scored 22–27 . . .

Oh yeah, you're a perfectionist! You take great care and pride in all that you do, and you expect the best from yourself. That part is great. But, when you make a mistake, you tend to beat yourself up and focus only on what you did wrong. You might even believe that others expect you to be "perfect."

Let's not confuse "perfectionist" with "high achiever." High achievers set high but realistic goals for themselves and go after them. They tend to bounce back from disappointments. But perfectionists can live in almost constant fear of making mistakes, and this can be a huge time drain. Have you ever been studying for a test, only to realize you've spent a huge percentage of your time worrying instead of studying?

So, how can you be a high achiever without being a perfectionist? Make a point to be forgiving of mistakes—yours and other people's. If you know you studied the best you could, decide that you'll feel great about a 92% on a test. I mean, GREAT. And what if you tried saying things to yourself that are comforting and encouraging instead of critical? As a rule, don't say anything to yourself that you wouldn't say to someone else!

I bet you've heard the expression "Nobody's perfect." Recite this over and over in your head until you believe it. Ironically, your grades will go UP and your creative juices will flow when you stop making so many rules for yourself. Don't worry if it takes a little time to change your patterns; any progress you make will benefit you in the long run!

Chapter 24

Puppies, Please?

Solving Quadratic Equations by Factoring

*I*magine walking up to your house, and your front door is being guarded by a very large, scary-looking pit bull. Yikes. But what if you had the power to magically turn that big dog into two little puppies. Would you do it? Of course you would!

In the next three chapters, I'll show you a few different ways to solve **quadratic equations**. For this first method—solving by factoring—the puppies are going to help.

Ring Ring What's It Called?

A **quadratic equation** is an equation in one variable that can be rewritten in the form $ax^2 + bx + c = 0$ (this is **standard form**), where a, b, and c are all real numbers and $a \neq 0$. These are all quadratic equations:

$$x^2 = 25 \qquad 3x^2 - 5x - 2 = 0 \qquad y^2 = -4 \qquad \frac{z}{2} = z^2 + 1$$

The solutions to these problems are often called the **roots*** of the equation. Quadratic equations can have two, one, or even *no* real solutions. For example, $y^2 = -4$ has no real solutions.[†] However, the equations below are *not* quadratic:

$$y^3 + y = 5 \qquad 3x + 2x - 5 = 0 \qquad \frac{x^2}{x} + 3x = 6$$

. . . .

* I won't use the word *root* very often, but it means the same thing as "solution."
[†] If a quadratic equation has <u>no real solutions</u>, that means it actually has two complex or imaginary solutions, which you'll learn about in Algebra II. (Yes, I said *imaginary!*)

I bet you already know how to solve some quadratic equations, like this:

$$x^2 = 25$$

You probably know that $x = 5$ is a solution to this, and maybe you even realize that it has a second solution, $x = -5$. It sure would be nice if all quadratic equations were as easy as $x^2 = 25$, wouldn't it? And . . . now that we're done daydreaming, let's get back to reality (it's not so bad, I promise).

So, how the heck do we find the values of x that satisfy an equation like this, anyway?

$$x^2 + 12 = 7x$$

Hmm. I mean, sure, we could always just start plugging in numbers and hope that something works, but there must be a better way. And there is! First, you know how zero times anything equals zero? It might seem obvious to you, but it's super important for solving quadratic equations, so check it out.

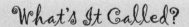

What's It Called?

Zero-Product Property of Real Numbers

For all real numbers a and b,

If $a = 0$ or $b = 0$, then $ab = 0$ (That's pretty easy, right?)

and the converse:

If $ab = 0$, then $a = 0$ or $b = 0$.

So if I tell you that $ab = 0$, then you will know that $a = 0$ or $b = 0$.*

. . . .

* It could also be true that a and b both equal zero.

Pretty simple rule, right? But values don't always come in neat 'n tidy packages like *a* or *b*. Sometimes they look more complicated, like $c - 4$. So if someone told us that $a(c - 4) = 0$, then by the same exact logic, we would know that either $a = 0$, or $c - 4 = 0$, right? In other words, we would *know* it must be true that either $a = 0$, or $c = 4$. Think about that for a second.

With me so far? Great! Let's say someone told us that $(c + 1)(c - 4) = 0$. We'd *know* that either $c + 1 = 0$, or $c - 4 = 0$. In other words, we'd *know* that either $c = -1$, or $c = 4$, right? Same logic. Now, can you please tell me the values of *x* that make this a true statement?

$$(x + 1)(x - 4) = 0$$

Yes, you can! This is a true statement if **$x = -1$** or **$x = 4$**. Guess what? You've just solved your first quadratic equation. You might say, "But that didn't look like a quadratic equation!" Multiply out the left side,* and you'll see that this was the equation below *in disguise*:

$$x^2 - 3x - 4 = 0$$

Try plugging in $x = -1$, and you'll see that we do get a true statement. The same is true if we plug in $x = 4$. See? It works! So when you see something like **$(x + 1)(x - 4) = 0$**, where two parentheses multiply together to equal zero, the Zero-Product property tells us we can find the solutions just by setting up these two mini equations, each equaling zero:

$$x + 1 = 0 \text{ and } x - 4 = 0$$

And those equations are short and sweet—just how we like 'em.

⌒‿‿‿‿‿‿◌

* See Chapter 21 to brush up on multiplying out binomials with FOIL.

✏️ Doing the Math

For each equation, do two things: Find the values that make the equation a true statement, and then rewrite the equation in standard form. I'll do the first one for you.

1. $(2x + 15)(2x - 3) = 0$

<u>Working out the solution</u>: Let's find the values of x that make each parentheses equal to zero! We'll solve $2x + 15 = 0$ by subtracting 15 from both sides and dividing both sides by 2, and we get $x = -\frac{15}{2}$. Similarly, $2x - 3 = 0$ will be true when $x = \frac{3}{2}$. So the equation is satisfied by the two values $x = -\frac{15}{2}, \frac{3}{2}$. For the next part, in order to rewrite the original equation in standard form, we just need to multiply it out and simplify. So, using FOIL, the left side becomes $(2x + 15)(2x - 3) = 4x^2 - 6x + 30x - 45 = 4x^2 + 24x - 45$. Our equation in standard form is $4x^2 + 24x - 45 = 0$.

Answer: $4x^2 + 24x - 45 = 0$ is satisfied by $x = -\frac{15}{2}$ or $\frac{3}{2}$.

2. $(x + 2)(x - 1) = 0$

3. $(y + 2)(y + 2) = 0$

4. $(3t - 1)(t - 4) = 0$

5. $(2x + 3)(3x - 5) = 0$

(Answers on p. 411)

QUICK NOTE As in problem 3 above, when there's only <u>one</u> value that satisfies a quadratic equation, it's called a "double root" or "double solution." It's like that one solution is pulling double duty, since it's all alone!

Enter the Puppies: Solving Quadratic Equations by Factoring

Oh, if only all quadratic equations were written like $(x - 3)(x - 4) = 0$, then we'd always know how to solve them, wouldn't we? The cold hard truth is, they're usually written more like $x^2 + 12 = 7x$. Yep, big, scary, pit-bull equations. But that's okay!

Remember the reverse FOIL technique we did in Chapter 22? We'll use it to *factor* $x^2 + 12 = 7x$, magically turning a scary pit bull into two puppy-like equations, $x - 3 = 0$ and $x - 4 = 0$, all cute and ready for us to solve. Nice! Here's the step by step:

Step By Step

Solving quadratic equations by factoring:

Step 1. Rewrite the equation so that all the "stuff" is on one side and x^2's coefficient is positive: $ax^2 + bx + c = 0$, $a > 0$. Also, factor out the GCF, if one exists.

Step 2. Next, use the reverse FOIL factoring technique from p. 328 or 332 (or *grouping*—see p. 338) to rewrite the equation into this "two puppy" format: (factor #1)(factor #2) = 0.

Step 3. Because of the Zero-Product property, you can now take each factor—each parenthetical expression—and set it equal to zero. Solve each of these easy mini equations, and those are your solutions!

Step 4. Plug the solutions into the *original* equation, and make sure they satisfy it. Done!

Solve $x^2 + 12 = 7x$.

Okay, so we're supposed to find the values of x that satisfy this. First, breathe. We've got a pit bull on our hands, but soon enough it'll become two little puppies.

Step 1. Subtracting $7x$ from both sides and putting it in descending order, we get our equation in standard form: $x^2 - 7x + 12 = 0$.

Step 2. It's time to turn this pit bull into puppies! We'll use the reverse FOIL method from Chapter 22 to factor the left side. Because x^2's coefficient is 1, all we have to do is find the pair of numbers whose product is 12 and whose sum is -7. The winning combo is -3 & -4, and we get $(x - 3)(x - 4)$. Try multiplying it out, and you'll see that we did it right! So, $x^2 - 7x + 12 = 0$ has become:

$$(x - 3)(x - 4) = 0$$

Step 3. Ah, the puppies. So cute. Using the Zero-Product property from p. 354, we know that if either of the two factors, $(x - 3)$ or $(x - 4)$, equals zero, then the whole left side equals zero, which would mean that the equation was satisfied. So our little mini equations are $x - 3 = 0$ and $x - 4 = 0$, which means our solution is $x = 3, 4$.

Step 4. Let's check to see if they really work:

$$
\begin{aligned}
&\text{Plugging in } x = 3 & &\text{Plugging in } x = 4 \\
&x^2 + 12 = 7x & &x^2 + 12 = 7x \\
&\rightarrow (3)^2 + 12 \stackrel{?}{=} 7(3) & &\rightarrow (4)^2 + 12 \stackrel{?}{=} 7(4) \\
&\rightarrow 9 + 12 \stackrel{?}{=} 21 & &\rightarrow 16 + 12 \stackrel{?}{=} 28 \\
&\rightarrow 21 = 21 \text{ Yep!} & &\rightarrow 28 = 28 \text{ Yep!}
\end{aligned}
$$

Yep, both values of x satisfy the equation. Done!

Answer: $x = 3, 4$

QUICK NOTE   Some teachers prefer answers to be left in "solution set" form. So our above solution could also be written like this: x ∈ **{3, 4}**. It's customary to write the numbers in order, lowest to highest. (I like to write them in that order even when I don't use set notation.)

Watch Out!

Just because we have two parentheses multiplied together doesn't mean we can split it up into two mini equations. <u>The parentheses must multiply together to equal **zero**</u>. So:

$$(x - 2)(x + 3) = 0 \qquad \rightarrow \qquad x - 2 = 0 \text{ or } x + 3 = 0$$

BUT:

$(x - 2)(x + 3) = 6$ does NOT mean that $x - 2 = 6$ or $x + 3 = 6$. For this problem, we'd have to multiply everything out using FOIL, rewrite in standard form (we'd get $x^2 + x - 12 = 0$), and then factor it to get $(x + 4)(x - 3) \rightarrow$ **$x = -4, 3$**.

QUICK NOTE   Notice that the sign on the *solution* is always the opposite from the sign in its *parentheses*. So if $(x + 1) = 0$, then **$x = -1$**, and if $(x - 1) = 0$, then **$x = 1$**.

Find all values that satisfy the following equations by factoring the big equations into two mini equations, using the Zero-Product property. I'll do the first one for you.

1. $-2x^2 + 13x = -7$

<u>Working out the solution</u>: Let's add 7 to both sides: $-2x^2 + 13x + 7 = 0$. It's much easier to factor when x^2's coefficient is positive, so we'll multiply both sides by -1: $(-1)(-2x^2 + 13x + 7) = (-1)(0) \rightarrow 2x^2 - 13x - 7 = 0$. Time to make puppies! On p. 333, we factored the left side of this into $(2x + 1)(x - 7)$, so our mini equations are $2x + 1 = 0$ and $x - 7 = 0$. We can solve $2x + 1 = 0$ by subtracting 1 from both sides and then dividing both sides by 2, which gives us $x = -\dfrac{1}{2}$. And $x - 7 = 0$ is true when $x = 7$. Let's check our work:

$$\underline{\text{Plugging in } x = -\tfrac{1}{2}}$$
$$-2x^2 + 13x = -7$$
$$\rightarrow -2\left(-\tfrac{1}{2}\right)^2 + 13\left(-\tfrac{1}{2}\right) \overset{?}{=} -7$$
$$\rightarrow -2\left(\tfrac{1}{4}\right) - \tfrac{13}{2} \overset{?}{=} -7$$
$$\rightarrow -\tfrac{1}{2} - \tfrac{13}{2} \overset{?}{=} -7$$
$$\rightarrow -\tfrac{14}{2} \overset{?}{=} -7$$
$$\rightarrow -7 = -7 \text{ Yep!}$$

$$\underline{\text{Plugging in } x = 7}$$
$$-2x^2 + 13x = -7$$
$$\rightarrow -2\,(7)^2 + 13\,(7) \overset{?}{=} -7$$
(Let's avoid big numbers, and use reverse distribution to factor out 7)
$$\rightarrow 7\big[-2(7) + 13\big] \overset{?}{=} -7$$
$$\rightarrow 7\big[-14 + 13\big] \overset{?}{=} -7$$
$$\rightarrow 7\,[-1] \overset{?}{=} -7$$
$$\rightarrow -7 = -7 \text{ Yep!}$$

Answer: $x = -\dfrac{1}{2}, 7$

2. $x^2 + 6x - 7 = 0$ 5. $4y^2 - 24 = 20y$

3. $x^2 - 6x = 7$ 6. $(x - 2)(x + 3) = 14$

4. $10w^2 = 5w + 5$ 7. $-4h^2 - 3h = -1$ *(Watch those negative signs!)*

(Answers on p. 411)

Reality Math

Decorating Diva

So, you're on the decorating committee for your school, and the committee has decided to hang a beautiful tapestry on the front of the main building, honoring the school's new music program. You've been entrusted with the task of finding the material to border it! The tapestry right now is **5 feet tall** and **7 feet wide**, so its area is **35 square feet**. You found some beautiful glittery material for a big border to go around the entire thing, and you found these great 1-foot-tall music notes to put on the border! But this special material is expensive, and due to budget constraints, you're only allowed to buy **45 square feet of it**—that would be the new border's total *area*. Sounds like plenty, but it's hard to tell. I mean, if the border has a uniform thickness all the way around, how thick can the border end up being? Would the music notes fit? It's always nice to know how something will look before spending money on it, after all! Let's sketch it, knowing that the total area will be 35 + 45 = **80 square feet**, and let's label the width of the border "*x*."

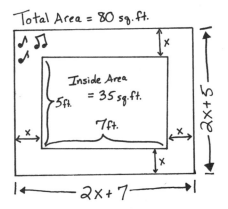

We know the total area is 80 square feet. Looking at the diagram, now let's write the total area *in terms of x*. From our diagram, the total length is **2x + 7**, and the total height is **2x + 5**. For this rectangle, length × height = area, so we can write this true statement involving *x*. Then we'll just solve for *x*!

$$(2x + 7)(2x + 5) = 80$$

We'll multiply out the left side with FOIL, simplify, and make sure 0 is on one side:

$$(2x + 7)(2x + 5) = 80$$

$$4x^2 + 10x + 14x + 35 = 80$$

$$\rightarrow \quad 4x^2 + 24x - 45 = 0$$

Time to make puppies! In problem 1 on p. 356, we saw that this equals:

$$(2x + 15)(2x - 3) = 0*$$

$$\rightarrow \quad x = -\frac{15}{2}, \frac{3}{2}$$

Only the positive solution, $x = \frac{3}{2}$, makes "sense" for the width of our border, so we'll only check that one:[†]

$$(2x + 7)(2x + 5) = 80$$

$$(2\left(\frac{3}{2}\right) + 7)(2\left(\frac{3}{2}\right) + 5) \overset{?}{=} 80$$

$$\rightarrow \quad (3 + 7)(3 + 5) \overset{?}{=} 80$$

$$\rightarrow \quad (10)(8) \overset{?}{=} 80 \text{ Yep!}$$

The borders will be $\frac{3}{2} = 1\frac{1}{2}$ **feet** all the way around. Smaller than you thought, huh? Well, at least we know it will work and the 1-foot-tall music notes will fit. Fabulous!

. . . .

* To factor this from scratch, we could have used grouping (p. 338), where $ac = -180$ and $b = 24$. Easy factor pairs of 180 are 6 & 30, 3 & 60, 9 & 20, and 2 & 90. And since $30 - 6 = 24$, we'd split the middle term, $24x$, into $30x - 6x$, and proceed with grouping!
† In quadratic equations, you'll find that often, only one of the solutions will answer a real-world scenario. Just because our real-world solution satisfies an equation doesn't mean that some other wacky number can't also satisfy it, right?

Get in the Driver's Seat!

Tired of solving equations? Let's take a break from all this "solving" and get in the driver's seat. See, teachers have it good. Instead of doing homework, they get to make up problems for students to solve. A teacher might think to herself, "Hmm, I want my students to solve a quadratic equation whose solutions are, oh, I don't know, how about $x = 1$ and -3?" And then all she has to do is write $(x-1)(x+3) = 0$. See? That is guaranteed to have the two solutions $x = 1$ and -3.

But, of course, the teacher doesn't leave it at that. She'll go ahead and multiply it out: $(x-1)(x+3) = 0 \rightarrow x^2 + 3x - 1x - 3 = 0 \rightarrow \boldsymbol{x^2 + 2x - 3 = 0}$. And that's what ends up on your test, delivered for you to solve. Well, power to the people, I say! It's *our* turn to do the driving.

Doing the Math

Now you're the teacher. Create the quadratic equation in standard form (and *without* fractions in it) for someone else to solve, <u>which has the solutions indicated</u>. I'll do the first one for you.

1. $x = -\dfrac{4}{5}, 9$

<u>Working out the solution</u>: Okay, so we want to build an equation that has these solutions—let's figure out what each parentheses should have in it. We'll start with $x = 9$, which is easier. After all, $(\boldsymbol{x-9}) = \boldsymbol{0}$ when $x = 9$, right? How about the other parentheses? Hmm, it's certainly true that $\left(x + \dfrac{4}{5}\right) = 0$ when $x = -\dfrac{4}{5}$, but let's get rid of the fraction by multiplying both sides by 5, and we get $(5)\left(x + \dfrac{4}{5}\right) = (5)0 \rightarrow \boldsymbol{5x + 4 = 0}$.

Yep, that's true when $x = -\dfrac{4}{5}$.* Now our full equation looks

---

* After all, we did the same thing to both sides of the equation, so the solution better not have changed.

like this: $(5x + 4)(x - 9) = 0$. And, it has the solutions we want it to! Time to multiply it out, and it'll be ready for some poor student to solve: $5x^2 - 45x + 4x - 36 = 0 \rightarrow$ $5x^2 - 41x - 36 = 0$. We've totally created a problem that has the solution $x = -\dfrac{4}{5}$, 9. And the best part is, we don't have to solve it.

Answer: $5x^2 - 41x - 36 = 0$ has the solutions $x = -\dfrac{4}{5}$, 9.

2. $x = -1, 5$ 5. $x = -1, -5$

3. $x = 1, -5$ 6. $x = \dfrac{2}{3}, -\dfrac{1}{5}$

4. $x = 1, 5$ 7. Make up your own!

(Answers on p. 411)

What's the Deal?

Not every quadratic equation can be solved by factoring with reverse FOIL. That's because many quadratic equations' solutions aren't even rational! That's right, the solutions can't even be written as a fraction. Their solutions might look like $x = 1 \pm \sqrt{3}$ or something else wacky and irrational. I'll show you a few tricks for handling those in the next couple of chapters.

Takeaway Tips

Quadratic equations can be written in the form $ax^2 + bx + c = 0$ (this is standard form), where a, b, and c are all real numbers and $a \neq 0$. They can have two solutions, one "double" solution, or no real solutions.

The Zero-Product property says that zero times anything equals zero, and also that if two factors multiply to equal zero, then one of them must itself *be* zero!

 To solve a quadratic equation by factoring, use reverse FOIL to write a standard form equation as (factor #1)(factor #2) = 0. Then, because of the Zero-Product property, each parenthetical factor (aka, what's in the parentheses) can be set equal to zero for two mini equations to solve.

 Always check your answers by plugging the solutions into the original equation.

Attitude Adjustment

Do you wish you had the magic power to *feel better* when life seems totally unfair? Well guess what—you can! You might be surprised how good it feels to take control of your own attitude with these tips. . . .

We Feel ⇨ ⇨ ⇨ ⇨ ✳ We Can Feel

Powerless, unsatisfied, and/or that things aren't fair

Powerful and grateful

> *Your locker is next to your nemesis, and now you have to see her every day!*

✳ **Attitude Adjustment!** There will always be people you don't want to deal with in life; this is actually good practice. I mean, what if someday, at your dream job, your coworker is a horrible person . . . who happens to be your boss's favorite niece? The key is to find common ground with that person, and keep it simple! Compliment her shoes or her performance in debate class—something you honestly admire. You're a smart cookie; you can find something! Keep conversations short, sweet, and sincere. It might not be as bad as you think.

> *You hate the "health" food you are being made to eat for dinner. Would fries and diet soda be so bad?*

Attitude Adjustment! Greasy fries and artificially sweetened soda are fine . . . as long as you don't mind acne and unhealthy weight gain. Think of your food as fuel. You are what you eat. Still not feeling grateful for the food you eat at home? Do an internet search for "starving babies in Africa." You'll get over that feeling pretty fast.

> You share a room, and your sister is totally messy. Every time you walk through, you feel angry.

Attitude Adjustment! This is pretty common, and it could happen throughout your life with college roommates and then maybe even your husband someday! You can't control other people's behavior, but you can control your own. Realize that you have a *choice*. You can choose to help your sister clean the room or make her a box for organizing her things. You might say, "She doesn't deserve my help!" But who cares? That doesn't change the fact that it's your choice, plain and simple. Realizing that you are *choosing* whether or not to help will make you feel more powerful; in fact, it will make you realize the power you already have. Besides, there's probably something you do that really annoys her, too.

> Again, it's 10 P.M. and you haven't even started your biggest homework assignment! Even if you promise yourself that you won't procrastinate, it still happens all the time. The whole thing is very stressful.

☑ **Attitude Adjustment!** For many people, procrastination is like a drug. The time spent avoiding a specific task is addictive. You probably have a few favorite distractions, and it's so easy to put off homework, chores, or other obligations. But you CAN regain control of your time: Remind yourself of the anxiety you feel when you procrastinate, and try making a <u>list of goals</u> for the day. It really works for me!

I have to make a paper list; if it's on my computer, I just ignore it for some reason. Plus, it feels really good to physically check off (or scratch out!) the stuff I've finished. If it's a huge assignment you're putting off, do at least *some* small piece of it each day. Something is better than nothing.

(See "50 Things This Week" on p. 89)

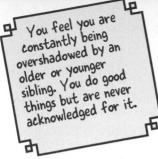

You feel you are constantly being overshadowed by an older or younger sibling. You do good things but are never acknowledged for it.

👑 **Attitude Adjustment!** Ask yourself this question honestly: "What is my motivation for getting good grades or excelling in other areas?" Is it really just to get your parents' attention? Remember that your parents are just people with their own insecurities and problems. Parents love each of their children equally, but no one is perfect. They might not realize that you crave more attention and validation than they are providing. If you want to, you can speak calmly and honestly with them about it. But try focusing on being grateful for the fact that your motivation for excelling is truly for *you* and *your* goals, and the life *you* will continue to lead, even after you move out of your parents' house in a few years. Of course, it feels great to get a pat on the back . . . but get good at patting *yourself* on the back by feeling proud of yourself for the hard work you do. Think of other people's validation as a bonus, nothing more. It'll make you a more independent, self-sufficient young woman for years to come!

Your parents forbid you to go to the biggest party of the year.

🌸 **Attitude Adjustment!** If your parents don't want you to attend the party, it's probably for a better reason than you think—even if they don't do a good job of describing it. Your parents are your parents, no matter what. And contrary to what you think, they have been in the same situations you find yourself in, and they're looking out for your best interests. It could be that the parents of the hosts are never around, and/or illegal, dangerous stuff could be present, like drugs. As much as it sucks, the truth is, you're not too young to make huge, life-changing mistakes. So instead of stressing about missing that one party (which in 5 years you'll have no memory of), invite your friends over and have your own get-together! By the way, although it's common to envy the kids whose parents don't care *what* they do . . . the truth is, those kids don't usually feel very loved.

Somehow, you are always late for the school bus. If your siblings didn't take such long showers, you're sure this wouldn't be an issue.

🏫 **Attitude Adjustment!** Realize that you have a choice; you could plan to get up 20 minutes early for the next week and see how much of a difference it makes. You'll be the first one in the shower, and you'll arrive at school on time and looking good. If you don't want to get up early, that's totally fine. But recognize that you actually have a choice, and YOU are the one who is making it.

All your friends have their belly buttons pierced and you want to do it, too. But your mom says no: "You're too young and it sends the wrong message to guys."

🐱 **Attitude Adjustment!** Well, she might have a point. Besides, belly-button rings can often be painful and annoying. You want to be attractive and alluring? Find that through confidence. A winning smile that reveals true confidence will always be more attractive than a belly-button ring on a girl consumed with insecurities. So instead of following the "belly-button ring" crowd, stand out on your own by getting good grades, excelling in sports or an extracurricular activity, and finding the confidence that comes from success and achievement. And if you still want your belly button pierced when you're older (even after half your friends' belly buttons got infected and they took theirs out for good), you totally can.

Have more examples of this? Send them to me at: share@danicamckellar.com

And here's a little mantra that always helps: "Thank you." I know it sounds simple, but it works! The next time you're super mad at someone or feeling out of control of your life, stop yourself, take a deep breath, go to a different room, and start silently saying "thank you" to yourself over and over until you think of something that you're actually thankful for. It could be your health, the power you do have, your pet, anything. Practice closing your eyes and actually *feeling* grateful. If you've ever thought about being an actress, this is great practice at being able to access a particular emotion at will. Plus, it feels great. And then think about *your* part in the whole situation—the part you can control. You can even journal about it. It'll save a lot of fights from getting big, and it will make everyone—especially *you*—feel better, so you can go back to being a fabulous, contributory young woman! For more ideas on how to change your mood, and tips on expressing your feelings using writing and journaling, check out pp. 113–15 and pp. 122–23. It's not always easy to adjust our attitude, but it sure does feel good.

Chapter 25

The Lemonade Stand

Solving Quadratic Equations by Completing the Square

"When life hands you lemons, make lemonade." Take something sour and make it sweet—what a nice sentiment. Of course, there is a flaw in the logic. I mean, if we already *had* the sugar (or better yet, honey*) needed to make the lemonade, then getting handed lemons would never have seemed like a bad thing in the first place, right? We would have been like, "Yay, lemons!" But . . . I still like the sentiment.

Sometimes in *math* we get handed lemons, too. In this chapter, we'll learn a technique for solving quadratic equations called "completing the square," which you'll see is pretty, um, lemony. Yes, we'll be handed lemons and gosh darn it, I'm going to show you how to make quadratic lemonade.

That just didn't sound as tasty as I was hoping.

Moving Toward "Completing the Square" . . .

Remember in the last chapter when we did this easy problem, $x^2 = 25$, and we knew the solution was $x = -5, 5$?

. . . .

* To make lemonade with honey, use a little warm/hot water (be careful!) to dissolve the honey and *then* mix it with the fresh-squeezed lemon juice and ice. Yum!

Here's what is actually going on:

$$x^2 = 25$$

(taking the square root of both sides)

$$\longrightarrow \sqrt{x^2} = \sqrt{25}$$

(We know the right side equals 5.)

$$\longrightarrow \sqrt{x^2} = 5$$

(Hmm, looks like x could be 5 or -5 to satisfy this.)

$$\longrightarrow x = \pm 5$$

$$\longrightarrow \textbf{x = -5 or 5}$$

QUICK NOTE   The symbol $\pm$ is read out loud like this: "plus or minus." Basically, it's a shortcut for writing two expressions in one compact form. So these all mean the same thing:

$$x = \pm 4 \quad x = 4 \text{ or } x = -4 \quad x \text{ equals plus or minus } 4$$

Just remember this: **Whenever we take the square root of both sides of an equation with a variable on one side, we should stick $\pm$ on the other side.** Now we get *two* equations instead of one! (See the What's the Deal? on pp. 381–82 for more on this.)

So, $x^2 = 25$ was pretty easy to solve, right? How about this:

$$(w - 1)^2 = 25$$

Let's use the same method: we'll take the square root of both sides and remember to use $\pm$, because we know that $(w - 1)$ could be positive or negative, just like x from the previous example:

$$\sqrt{(w - 1)^2} = \sqrt{25}$$

$$\rightarrow \sqrt{(w - 1)^2} = 5$$

$$\rightarrow (w - 1) = \pm 5$$

which is the same as:

$$w - 1 = -5 \quad \text{or} \quad w - 1 = 5$$

And solving these two mini equations is easy! We end up with the solution **w = –4, 6**. Try plugging those values into the original equation, $(w - 1)^2 = 25$, and you'll see that they both work. Nice.

QUICK NOTE   Often we'll end up with a square root symbol in the answer, so instead of something like ±5, we'd get something more like ±√5. Everything else works exactly the same way.

Doing the Math

Solve these equations. Then put the equation in standard form: $ax^2 + bx + c = 0$. I'll do the first one for you.

1. $(w - 2)^2 = 3$

<u>Working out the solution</u>: To solve for w, we take the square root of both sides and get $\sqrt{(w - 2)^2} = \sqrt{3}$. Unfortunately, 3 isn't a perfect square, so we have to leave the right side of the equation as is. Simplifying the left side, and knowing we'll have to stick ± on the right: $(w - 2) = \pm\sqrt{3}$. This becomes two mini equations: **(w – 2) = –√3** or **(w – 2) = √3**. To solve them, we can drop the parentheses, add 2 to both sides on each, and we end up with **w = 2 –√3** for the first solution, and **w = 2 + √3** for the second one. Finally, let's multiply out the original equation and put it in standard form: $(w - 2)^2 = 3$ → $w^2 - 2w - 2w + 4 = 3$ → **w² – 4w + 1 = 0**.

Answer: w² – 4w + 1 = 0 has the solution: w = 2 – √3, 2 + √3.

2. $(y - 1)^2 = 16$

3. $(t + 3)^2 = 4$

4. $(x - 3)^2 = 5$

(Answers on p. 411)

Back to the Lemons

We were able to solve each of the problems in this Doing the Math because we had *something squared* on the left, like $(w - 2)^2$, and we could conveniently take the square root of both sides to help isolate the variable. But we won't always be handed lemonade. No, sometimes we'll be handed an equation that simply isn't in a form where we can just take the square root of both sides (and it can't be factored, either). Totally lemons:

$$x^2 + 6x = -2$$

But we'll take the ingredients we're given and make something sweet out of them: <u>We'll make the left side into a perfect square, so that we **can** take the square root of both sides!</u> It's called "completing the square."

Completing the Square

Our goal in *completing the square* is to take an equation that looks something like $x^2 + 6x = -2$ and rewrite it so it looks more like $(x + d)^2 = e$, where d and e are both constants—just plain 'ol numbers. If we do that, then x can be isolated by taking the square root of both sides, like we've just been doing, right?

So the question is, how the heck are we supposed to make an equation look like $(x + d)^2 = e$? I'm glad you asked.

Just for the fun of it, let's see what the left side, $(x + d)^2$, looks like when it's all multiplied out.*

$$(x + d)^2 = (x + d)(x + d) = x^2 + 2dx + d^2$$

Knowing this, then let's make our <u>new goal</u> to get our equations to look like this:

$$x^2 + 2dx + d^2 = e$$

...because then we know we'll be fine, since this is the same as $(x + d)^2 = e$, right? I want you to read the last couple of lines again until they make total sense. Don't let all the letters get you down!

. . . .

* To brush up on multiplying binomials with FOIL, see p. 310.

Hmm, let's see how close $x^2 + 6x = -2$ is to the form
$x^2 + 2dx + d^2 = e$:

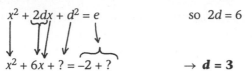

$$x^2 + 2dx + d^2 = e \qquad\qquad \text{so } 2d = 6$$

$$x^2 + 6x + \,? = -2 + \,? \qquad\qquad \rightarrow \boldsymbol{d = 3}$$

Well, the x^2 part looks pretty good. And our $6x$ could be the $2dx$ part, right? All we would need is for $2d$ to equal 6. In other words, $d = 3$. Okay, fine, then let's say $d = 3$. Well, we're still missing the d^2 term, which in this case should be 9, right? Oh, but we *can* have a 9 on the left. We'll just add 9 to *both* sides!

$$x^2 + 6x + \,? = -2 + \,?$$

$$x^2 + 6x + \mathbf{9} = -2 + \mathbf{9}$$

$$x^2 + 6x + 9 = 7$$

And now we're in the form $x^2 + 2dx + d^2 = e$, and $e = 7$. Great! Wait, why did we do this again? Oh yeah—because now the left side can be written as a squared parentheses.

$$\underbrace{x^2 + 6x + 9}_{} = 7$$
$$\overbrace{(x + 3)^2}^{} = 7$$

Yes, we knew the left side would become a perfect square, because that's exactly how we planned it.* We've transformed the equation into a totally equivalent statement, but now we have a <u>perfect square</u> on the left side. We've *completed the square*. Phew! That was the hard part. Isn't that crazy? This is the same equation as before, only now it has this nice lemonade-like quality. Much friendlier and easier to, uh, swallow.

So now that we have our equation written as $(x + 3)^2 = 7$, we can take the square root of both sides to finish it off, which we'll do later. This was pretty advanced stuff, so don't feel bad if you need to read the last couple of pages a few times!

. . . .

* A perfect time to channel your inner villain: *Ah yes, everything is working out according to my pah-lan . . .* [insert evil laugh here].

QUICK NOTE Notice how adding the square number, **9**, to both sides was the *key* to writing the left side as a perfect square trinomial. If we can figure out this "magic" square number, we're set.

Shortcut Alert

Finding the magic square number: Look at the coefficients!

As long as the coefficient for x^2 is 1, just <u>take the coefficient on x, divide it by 2, and then square it</u>. That's the magic square number! Divide by 2 and then square it. Got it? Good.

For example, let's say we want to complete the square for $x^2 + 8x = 1$. In this case, x's coefficient is 8, so we <u>divide it by 2</u> and get **4**, and then we <u>square</u> 4 to get **16**. That's our magic! So, we'd add **16** to both sides in order to complete the square.

For the equation $x^2 - 3x = 5$, x's coefficient is –3. We'd divide by 2 and get $-\frac{3}{2}$, and then square it: $\left(-\frac{3}{2}\right)^2 = \frac{9}{4}$. So we'd add $\frac{9}{4}$ to both sides. This stuff really works!

Step By Step

Solving quadratic equations by "completing the square":

Step 1. Put the equation into the form $ax^2 + bx = number$, where $a \neq 0$. If x^2's coefficient, a, isn't equal to 1, then <u>get rid of it</u> by dividing both sides by a (or multiplying both sides by $\frac{1}{a}$).

Step 2. Write blank spots on both sides where you'll put the "magic" square number.

Step 3. What's the magic missing square number? Take x's coefficient, <u>divide it by 2</u> (this is d, which eventually will go inside the parentheses, and be sure to notice if it's negative), <u>and then square it</u>. That will be the magic square number you want, which we'll call d^2. Add that new square number to *both* sides, to keep the scales balanced.

Step 4. Now that we have a perfect square trinomial on the left, rewrite it as a squared parentheses. So if we had something like $x^2 + 2dx + d^2 =$ *some number*, now our equation is $(x + d)^2 =$ *some number.* Remember, sometimes d will be negative. It's the magic square number right before you squared it! It's always a good idea to multiply out the parentheses just to make sure it's right.

Step 5. If you're asked to actually find the solutions, then take the square root of both sides of our lemonade-like equation, be sure to stick ± on the right side, and solve your two mini equations. Done!

And... Step By Step In Action
Action!

Rewrite the equation $3x^2 - 12x - 18 = 0$ into the form $(x + d)^2 =$ number.

Step 1. Let's get rid of x^2's coefficient by multiplying both sides by $\frac{1}{3}$:

$$\frac{1}{3}(3x^2 - 12x - 18) = \frac{1}{3}(0) \;\rightarrow$$

$$\frac{3x^2}{3} - \frac{12x}{3} - \frac{18}{3} = \frac{0}{3} \;\rightarrow$$

$$x^2 - 4x - 6 = 0$$

Adding 6 to both sides, we get $x^2 - 4x = 6$.

Steps 2 and 3. Now, we need to find the new "magic" square number to add to both sides.

$$x^2 - 4x + \_ = 6 + \_$$

Let's look at x's coefficient: −4. We're supposed to divide this by 2 and then square it. If we divide −4 by 2, we get **−2** (this is d), and then squaring it, we get **4**. This is our new "magic" square number! Let's add it to both sides:

$$\underline{x^2 - 4x + 4} = 6 + 4$$

Step 4. Now we can write the *left side* as a squared parentheses. Inside the parentheses, we use the magic square number *before* it was squared, in this case, **−2**:

$$(x-2)^2 = 10$$

Let's use FOIL* to multiply out the left side and make sure we get $x^2 - 4x + 4$: $(x - 2)^2 = (x - 2)(x - 2) = x^2 - 2x - 2x + 4 = \mathbf{x^2 - 4x + 4}$. Yep!

Answer: $3x^2 - 12x - 18 = 0$ can be rewritten as $\mathbf{(x - 2)^2 = 10}$.

Yep, those equations look pretty different, but they are satisfied by the *same solutions*.

Watch Out!

It's very easy to make a mistake at the stage where we wrote $(x - 2)^2 = 10$; for instance, we might have written $(x + 2)^2 = 10$ or even $(x + 4)^2 = 10$. I highly recommend quickly multiplying out the parentheses to make sure you got the correct d.

Take Two: Another Example

Complete the square $x^2 - x - 5 = 0$.

Step 1. The x^2 term already has a coefficient of 1, so we'll go ahead and add 5 to both sides to get it ready for the "magic" square number: $x^2 - x = 5$.

Step 2. Let's write some blank lines to show where we'll be adding the magic square number to both sides: $x^2 - x + \underline{\quad} = 5 + \underline{\quad}$.

Steps 3 and 4. To find the magic square number, we'll take x's coefficient, -1, and first divide it by 2 to get $\frac{-1}{2}$. Before we square it, let's make a mental note that $-\frac{1}{2}$ will be d, the number inside the parentheses. Squaring it, we get $\left(\frac{-1}{2}\right)^2 = \frac{1}{4}$, and that's our magic square number! Let's add it to both sides:

$$x^2 - x + \tfrac{1}{4} = 5 + \tfrac{1}{4}$$

· · · · · · · · · ·

* To review FOIL (First, Outside, Inside, Last), see p. 310.

To simplify the *right* side, we'll rewrite 5 as $\frac{20}{4}$, so we get $\frac{20}{4} + \frac{1}{4} = \frac{21}{4}$.

Our equation has become $x^2 - x + \frac{1}{4} = \frac{21}{4}$. Now we can rewrite the left side using $-\frac{1}{2}$ as the number inside the parentheses, just like we made a mental note to do:

$$\left(x - \tfrac{1}{2}\right)^2 = \tfrac{21}{4}$$

And voilà! We've completed the square and put it into a form where we can simply take the square root of both sides in order to solve for *x*. It's a good idea to quickly do a reality check to make sure the left side really does equal $x^2 - x + \frac{1}{4}$.

$$\left(x - \frac{1}{2}\right)^2 = \left(x - \frac{1}{2}\right)\left(x - \frac{1}{2}\right) = x^2 - \frac{1}{2}x - \frac{1}{2}x + \frac{1}{4} = x^2 - x + \frac{1}{4}$$

Yep! We did it correctly.

Answer: $x^2 - x - 5 = 0$ can be rewritten as $\left(x - \frac{1}{2}\right)^2 = \frac{21}{4}$

"*I* began to enjoy algebra when I realized it was like solving a mystery. I'd always enjoyed mysteries—and this was one I could solve on my own!" **Bailey, 15**

Doing the Math

Rewrite each equation in the form $(x + d)^2 =$ **number**. I'll do the first one for you.

1. $4x = 2x^2 - 6$

Working out the solution: First, let's flip everything so the x^2 term is on the left. It just makes me feel better: $2x^2 - 6 = 4x$. Now, let's add 6 to both sides and subtract $4x$ from both sides: $2x^2 - 4x = 6$. Oops, we still have to make sure x^2's coefficient is 1, so dividing both sides by 2 and leaving room for our magic square number, we get $x^2 - 2x +$ ___ $= 3 +$ ___. To find it, we'll take x's coefficient, -2, and first divide it by two: -1. We'll need that number in a moment; it'll go in the parentheses. For now, let's square it to get our magic square number: $(-1)^2 = 1$. Adding 1 to both sides, we get $x^2 - 2x + 1 = 3 + 1$. Great! Now the left side can be written as $(x - 1)^2$, just like we planned, and our equation now looks like this: $(x - 1)^2 = 4$.

Answer: $(x - 1)^2 = 4$

2. $x^2 + 8x = 1$

3. $3x^2 + 6x - 6 = 0$

4. $x^2 - 6x - 3 = 0$

5. $x^2 + x = 2$

(Answers on p. 411)

Solving Quadratics Using Completing the Square

Now that we've gotten some practice transforming equations so that the left side is a perfect square, let's finish the job by actually solving for x.

Solve $x^2 + 6x = -2$.

We started this on pp. 372–73; we completed the square and rewrote it as $(x + 3)^2 = 7$. Let's finish this up by taking the square root of both sides:

$$(x + 3)^2 = 7$$

$$\sqrt{(x + 3)^2} = \sqrt{7}$$

$$(x + 3) = \pm\sqrt{7}$$

So now we have two mini equations, $x + 3 = \sqrt{7}$ and $x + 3 = -\sqrt{7}$, right? Once we solve these, we'll have our solutions! To solve them, we just subtract 3 from both sides of each to isolate x and get $x = -3 + \sqrt{7}$ or $-3 - \sqrt{7}$. For reasons you'll understand later in Algebra II, we always put the square root part second.*

Answer:[†] $x = -3 - \sqrt{7}$ or $-3 + \sqrt{7}$

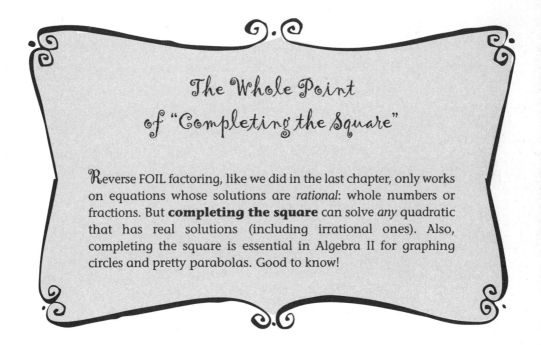

The Whole Point of "Completing the Square"

Reverse FOIL factoring, like we did in the last chapter, only works on equations whose solutions are *rational*: whole numbers or fractions. But **completing the square** can solve *any* quadratic that has real solutions (including irrational ones). Also, completing the square is essential in Algebra II for graphing circles and pretty parabolas. Good to know!

· · · · · · · · · ·

* When you learn about "conjugates," you'll see what I mean.
† For now, we won't worry about plugging these irrational numbers back into the original equation to check our answers; you'll do that in Algebra II.

Doing the Math

Solve for x by completing the square, first naming d^2. I'll do the first one for you.

1. $4x = 2x^2 - 6$

<u>Working out the solution</u>: On p. 378, we completed the square and rewrote this as $(x - 1)^2 = 4$ (and d^2 was 1). So, taking the square root of both sides, we get $\sqrt{(x - 1)^2} = \sqrt{4} \rightarrow$ $x - 1 = \pm 2$, which gives us two mini equations to solve: $x - 1 = 2$ and also $x - 1 = -2$. For each, we can isolate x by adding 1 to both sides. So we get $x = 3$ or -1.* It's easy to plug in each of these integer solutions and get true statements. Try it!

Answer: $d^2 = 1$; $x = -1, 3$

2. $x^2 + 10x + 9 = 0$ 4. $3x^2 - 12x = 6$

3. $x^2 + 2x = 4$ 5. $x^2 + 3x - 4 = 0$

(Answers on p. 411)

QUICK NOTE   If our equation is in the form $ax^2 + bx + c = 0$, then it turns out that d will always be equal to $\frac{b}{2a}$, so that means d^2 will be equal to $\left(\frac{b}{2a}\right)^2$. But I've found it much easier to use the steps as I showed you on pp. 374–75 to find d^2.

. . . .

* Since we got *rational* answers, we know that we also could have solved this equation by plain 'ol reverse FOIL factoring from Chapter 22. We'd get $(x + 1)(x - 3) = 0$.

What's the Deal?

Taking the Square Root and that Wacky ± Sign

What's the deal with the ± sign? Hmm, well we know that if we don't use it, we'll *lose* solutions, right? For example, if our original equation were $x^2 = 16$ and we took the square root of both sides without thinking about it, we might get $x = 4$. And we'd lose the perfectly good solution $x = -4$. Oops! But still, how does this ± symbol just pop up out of nowhere? Where does it come from?

So, we talked about square roots of *numbers* in Chapter 19, and you might have noticed that on p. 275, we said for all $a \geq 0$, $\sqrt{a^2} = a$. We only talked about <u>positive</u> values of a (and zero). Well, what happens if a is negative? *Then* what does $\sqrt{a^2}$ equal?

Because $(-4)^2 = 16$, this is true:

$$\sqrt{(-4)^2} = 4$$

When we square it and then take the square root, we go from -4 to 4, and that makes sense, right? So if what we stick under there is positive, it will stay positive, but if it's negative, it will become positive . . . sounds a lot like absolute value! So here's the full, grown-up version of what we saw on p. 275:

For *all* real numbers x,

$$\sqrt{x^2} = |x|$$

Don't let the brain freeze up—keep reading! All this means is that <u>no matter what x is</u>, positive or negative, it will become the <u>positive version of itself</u> after we square it and take the square root. Make sense?

Now I'm going to let you in on a little secret: This whole time, we've actually been *skipping a step* when we solve for x. Here's how it should really go:

$$x^2 = 16$$
$$\rightarrow \sqrt{x^2} = \sqrt{16}$$

by definition $\Big\downarrow \quad \downarrow$ by definition

$$\rightarrow |x| = 4$$

Because x could be positive or negative, this says the *positive version* of x is equal to 4. And as you know from solving absolute value equations in

Chapter 7, this statement is satisfied whether *x* equals –4 or 4, which is why we have to stick on the ±:

$$|x| = 4$$
$$\rightarrow x = \pm 4$$

And there's our answer! (pant, pant)

Teachers don't usually include the absolute value step because it can look confusing, like, "Hey, where did that absolute value come from?" And some teachers might not even understand it themselves. But now *you've* seen behind the curtain; you've seen how the invisible hand of the absolute value forces us to add the ± sign.

Okay, this was totally advanced, but let me tell you something. Just by <u>reading</u> it (even if you didn't totally get it), you have become a more powerful mathematician. And you know what? Read it again sometime.

Takeaway Tips

 The goal in **completing the square** is to get our quadratic equation to look like $(x + d)^2 = $ **number**. Because $(x + d)^2$ is the same as $x^2 + 2dx + d^2$, we just find the "magic" square number needed, d^2, and add it to both sides.

As long as <u>x^2's</u> coefficient is 1, the magic square number is <u>x's</u> coefficient, divided by 2 (this will be d, which might be negative), and then squared (this will be d^2).

 Once you've completed the square, you can take the square root of both sides and be sure to stick ± on one side. Then solve the two resulting mini equations.

TESTIMONIAL

Dr. Amber Wheeler
(Wayne, NJ)
Before: Worrier and target of mean-spirited jealousy
Today: Fabulous doctor in NYC!

   I grew up in a suburb in northern New Jersey, where being biracial was a bit of a novelty, and most of my classmates had more money than we did. I couldn't wear trendy, expensive clothes to school (like the popular girls) or travel on spring break, but I had good friends. Besides, I didn't have time to worry about being rich or "popular" because I was too busy worrying about living up to the high expectations of my parents—and myself. My sister had done very well in high school, and my parents were so proud of her. I wanted them to be that proud of

> "I worried all the time-often to the point of illness."

me, too! So I worked really hard in math and every other subject, and I worried all the time—often to the point of illness. *What if my teacher doesn't like my essay? What if I don't do well on next week's test? What if I don't understand the next chapter? What if . . . ?*

   Of course, making myself sick with worry never helped me in math or anything else. Luckily, I had managed to befriend many people outside of my direct academic circle by running track and joining a range of different school organizations, so that showed me that there was more to life than worrying all the time about grades. It was an important lesson!

   When I was a senior in high school, I learned another lesson—a hard one. I was accepted early to a prestigious university, much to the jealousy of some of my classmates. I began to hear malicious whisperings that I only got in because of affirmative action; in other words, that I only got in because I was half-black. I was devastated. Weren't these girls my friends? Didn't they know how hard I'd worked to achieve my good grades? I didn't realize at the time that there would always be people ready to cut down those who had success, no matter how much work it had required.

My mom suggested that, rather than be confrontational, I continue to be cordial to those girls and simply keep my good grades and achievements *private*, only to be shared with my most trusted friends. It was great advice that works for me, and that I follow to this day. (I'm grateful, though, to have been inspired by public figures who made the opposite choice at some point in their lives: Powerful women who put their accomplishments on display to help encourage us all to fulfill our maximum potential!)

Today, I love being a doctor, and I wouldn't be here without math. Not only did the college premed program require calculus and several science classes, but as a doctor, diagnosing a patient's disease is truly like solving for the variable in an equation: People often come into the hospital with no idea of what's wrong with them, and it's my job to figure it out in an efficient manner so that they get better as soon as possible. My profession is also very social—groups of doctors from all types of specialties constantly interact with one another. There is rarely a dull moment!

Speaking of "social," I am a young woman attempting to be fashionable and social while living in a big city on a limited budget, so math really helps me there, too. I've found creative ways to make my money last and pay off my credit cards and student loans, which is so important.

By the way, worrying is something I still struggle with, but I've learned to see the bright side: That same side of my personality knows how to work hard and achieve my goals, and that's led me to a life that I really love.

Chapter 26

What You Can Learn at the Mall

Using the Quadratic Formula

$\mathcal{P}$icture this: You're at the mall, and you're in such a bad mood that you can't tell if you feel like shopping or not. This might sound unlikely, but it happens . . . in fact, it happened to a girl named Betty, and this is going to help you learn the quadratic formula.

The Quadratic Formula

The quadratic formula seems a little bit like magic at first. It's a formula that can find the solutions to *any* quadratic equation. First, we make sure the equation is in the form $ax^2 + bx + c = 0$, **where a, b, and c are real numbers and $a \neq 0$**. Then we plug the values for a, b, and c into the equation below, and presto! We get the solutions.

There aren't many guarantees in life, but if real solutions exist to a quadratic equation, then you are guaranteed to find them. And now, I am proud to introduce . . .

(Drum roll optional)

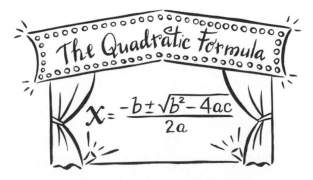

The Quadratic Formula

$$x = \frac{-b \pm \sqrt{b^2 - 4ac}}{2a}$$

And guess what? We have to memorize it. Yep, we really, really do. How can we do this? Well, for starters, by saying it out loud, several times. Try it now: "*x* equals negative *b*, plus or minus the square root of *b* squared, minus 4*ac*, ALL over 2*a*."

A common mistake is forgetting that the –*b* is also part of the big fraction, so that's why the "ALL over" is important when saying it out loud. But before we stress too much about how the heck we're supposed to memorize this big thing, I'd like to tell you a story about a girl named Betty . . . at the mall.

Betty was at the mall, and in such a negative mood that she couldn't decide if she wanted to enter the new radical store that just opened. It was known for having all sorts of crazy, radical designs on the clothes and shoes. She finally decided to go inside, and she immediately saw herself reflected in a huge double *square*-shaped mirror. It was Betty, squared! That made her smile a little. Suddenly, over her shoulder in the mirror, she spotted the 4 snobbiest girls in school, who seemed to always be in a negative mood: Anastasia, Abigail, Clarissa, and Claire. Betty quickly hid behind some clothes, not wanting to be seen. She didn't realize that two of the girls had crouched down low to look at some shoes. So when Betty came out from the clothes, she tripped right over them, spilling her lemonade ALL over 2 of them—Anastasia and Abigail!

The sales clerk started yelling, making all the girls laugh and go running out of the store together. Betty discovered that the girls weren't so bad after all, and she felt better all day. Looks like this whole wacky situation was the *solution* to fixing her bad mood!

Betty's Solution for a Bad Mood . . . and for memorizing the Quadratic Formula!

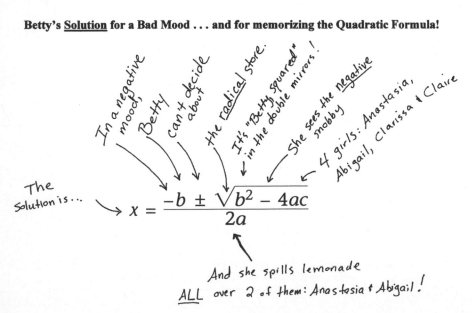

In a negative mood, Betty can't decide about the radical store. It's "Betty squared" in the double mirrors! She sees the negative snobby 4 girls: Anastasia, Abigail, Clarissa & Claire

The Solution is...

$$x = \frac{-b \pm \sqrt{b^2 - 4ac}}{2a}$$

And she spills lemonade ALL over 2 of them: Anastasia & Abigail!

. . . and they all became friends and lived happily ever after, learning to see other women first as allies, not enemies. They kicked butt in algebra class, too.

See? I told you that having a bad mood at the mall would help you learn the quadratic formula. Read the Betty story a couple more times, even read it out loud to your friends, pets, whoever—and then see if you can write down the quadratic formula without peeking. I bet you'll get pretty close on your first try! Do this for a week, just once a day, and you'll be in great shape.

Oh, and I should mention, some people like to "sing" this formula to the tune of "Pop Goes the Weasel."* (Don't worry; Betty helps whether you feel like singing or not.)

QUICK NOTE   The $\pm$ sign, which we saw in the previous chapter on p. 370, is a part of the quadratic formula. Remember, it indicates <u>two</u> different values. So, for example, an expression like $x = \dfrac{1 \pm 3}{2}$

means two statements: one with the plus sign, $x = \dfrac{1 + 3}{2}$, and one with the minus sign, $x = \dfrac{1 - 3}{2}$. And this is how we'd simplify them:

$$x = \frac{1 \pm 3}{2} \quad \rightarrow \quad x = \frac{1 + 3}{2} \text{ or } \frac{1 - 3}{2}$$

$$\rightarrow \quad x = \frac{4}{2} \text{ or } \frac{-2}{2} \quad \rightarrow \quad x = 2 \text{ or } -1$$

By the way, I usually like to list the lowest number first when writing the answer, so I'd write this solution as $x = -1, 2$.

Once again, say it with me slowly while thinking about Betty: "*x* equals negative *b*, plus or minus the square root of *b* squared minus 4*ac*, ALL over 2*a*." Good job!

• • • • • • • • • •

* Do an Internet search for "Pop Goes the Weasel quadratic formula" and you'll hear all sorts of people singing it. I saw the cutest two-year-old girl singing it. Yep, only two!

Step By Step

Solving quadratic equations using the quadratic formula:

Step 1. Rewrite the equation so it's in standard form: $ax^2 + bx + c = 0$.

Step 2. Identify a, b, and c for this particular equation; pay attention to negative signs.

Step 3. Write out the quadratic formula with a, b, and c. It's good practice, believe me.

Step 4. Substitute the a, b, and c values, using parentheses, and simplify.

Step 5. If the final solutions *don't* have square roots in them, plug them into the original equation and verify that you get a true statement.* Done!

QUICK NOTE   To be sure you use the correct a, b, and c values, make sure the equation is in standard form, and if you want, rewrite subtraction as "adding negatives": $3x^2 - 4x - 1 = 0 \rightarrow 3x^2 + (-4x) + (-1) = 0$; now it's easier to see that $a = 3$, $b = -4$, and $c = -1$.

And... Action! Step By Step In Action

Use the quadratic formula to find the values of x that satisfy this equation:

$$-2x^2 + 13x = -7$$

Let's follow the steps!

- - - - - - - - - -

* Checking *irrational* solutions is something you'll do in Algebra II.

Step 1. First, we rewrite it in standard form. We want all the "stuff" on the same side. So let's add 7 to both sides, and we get $-2x^2 + 13x + 7 = 0$.

Step 2. So, $a = -2$, $b = 13$, and $c = 7$.

Step 3. Let's write the quadratic formula: $x = \dfrac{-b \pm \sqrt{b^2 - 4ac}}{2a}$.

Step 4. Substituting our values for a, b, and c, and using parentheses, we get:

$$x = \frac{-(13) \pm \sqrt{(13)^2 - 4(-2)(7)}}{2(-2)}$$

Looks sort of scary! It's a good thing for the parentheses. They really help keep the negative signs organized. Simplifying, we get:

$$x = \frac{-13 \pm \sqrt{169 + 56}}{-4} \rightarrow x = \frac{-13 \pm \sqrt{225}}{-4}$$

You might already know that 225 is a perfect square. If not, you might notice that it's divisible by 5, so why not do a prime factorization* to see what's going on with its factors? $225 = 5 \times 45 = 5 \times 9 \times 5 = \mathbf{5 \times 5 \times 3 \times 3}$. Looks like we can rewrite this as $225 = 15 \times 15$. So that means $\sqrt{225} = 15$. Nice! So:

$$x = \frac{-13 \pm 15}{-4}$$

And I guess it's about time to separate this into its two solutions. The first one, using the *plus sign*, is $x = \dfrac{-13 + 15}{-4} = \dfrac{2}{-4} = -\dfrac{1}{2}$. And here's the second one, using the *minus sign*: $x = \dfrac{-13 - 15}{-4} = \dfrac{-28}{-4} = \dfrac{28}{4} = 7$.

Step 5. On p. 360, we solved this same equation with factoring and verified these solutions by plugging them in. So we're done!

Answer: $x = -\dfrac{1}{2}, 7$

Any time the solutions are rational (don't have any square roots in them), as in this previous example, it means it would have been possible to factor the equation with reverse FOIL like we did in Chapter 24. In fact, we would have gotten $(2x + 1)(x - 7) = 0$.

Factoring usually involves a lot less stuff to write down and simplify—and fewer chances for mistakes. Not every equation can be factored, but it sure is worth giving it the 'ol college try before succumbing to the quadratic

· · · · · · · · · ·

* To brush up on prime factorization, check out Chapter 1 in *Math Doesn't Suck*.

formula. I like to think of the quadratic formula as a dependable, trusty . . . last resort!

Watch Out!

Careless mistakes are easy to make. Be sure to start by writing the equation in *standard form*. If we looked at $8x^2 + 2x = 3$ too quickly, we might say $c = 3$. Oops! Also, don't skip the step of <u>writing out</u> what a, b, and c equal. You'd be amazed at how often that will save you from forgetting a negative sign. It's so much faster to write this stuff down to begin with, rather than try to find mistakes later. Trust me on this one.

Take Two: Another Example

Find the values of x that make this a true statement:

$$x^2 + 2x - 2 = 0$$

With such "nice" numbers, it's tempting to think we can factor this. But just try to set it up: $(x\ ?\ 1)(x\ ?\ 2)$. Darn it, there's no combo of + and – to make this equal $x^2 + 2x - 2$. So, it's the quadratic formula to the rescue!

Steps 1 and 2. It's already written in standard form, with $a = 1$, $b = 2$, and $c = -2$.

Step 3. Let's write the formula for practice: $x = \dfrac{-b \pm \sqrt{b^2 - 4ac}}{2a}$. (Remember Betty!)

Step 4. Plugging in our values, we get:

$$x = \frac{-(2) \pm \sqrt{(2)^2 - 4(1)(-2)}}{2(1)} \ \rightarrow \ x = \frac{-2 \pm \sqrt{4 + 8}}{2} \ \rightarrow \ x = \frac{-2 \pm \sqrt{12}}{2}$$

A perfect square factor, 4, is lurking inside that square root sign! Because $12 = 4 \times 3$, we can pull out 4, which becomes 2 on the outside of the radical sign:

$$x = \frac{-2 \pm \sqrt{4 \cdot 3}}{2} \quad \rightarrow \quad x = \frac{-2 \pm 2\sqrt{3}}{2}$$

...which means our solutions are $x = \frac{-2 + 2\sqrt{3}}{2}$ or $x = \frac{-2 - 2\sqrt{3}}{2}$. Notice we can *reduce* these fractions by factoring out a 2 on top and then canceling. Here's the first one:

$$x = \frac{-2 + 2\sqrt{3}}{2} \quad = \quad \frac{\mathbf{2}(-1 + \sqrt{3})}{\mathbf{2}} \quad = \quad -1 + \sqrt{3}$$

Similarly, the second solution reduces and becomes $-1 - \sqrt{3}$.

Step 5. Since our two answers are irrational, we won't worry about checking them. We'll get plenty of practice doing that in Algebra II, but for now, we're done!

Answer: $x = -1 - \sqrt{3}, \ -1 + \sqrt{3}$

Watch Out!

When reducing these fractions, make sure *every* term (on top and bottom) has the factor you want to cancel. For example, there is no way to reduce this further: $x = \frac{-2 \pm \sqrt{2}}{2}$. So the final answer should be $x = \frac{-2 - \sqrt{2}}{2}, \frac{-2 + \sqrt{2}}{2}$. It sure seems like we should be able to factor out a 2, but remember that the 2 inside the radical sign doesn't count.

Let's practice!

Doing the Math

Solve these using the quadratic formula. Be sure to write out the formula each time. I'll do the first one for you.

1. $\frac{x^2}{9} - \frac{x}{3} = 1$

Working out the solution: Let's get rid of the fractions by multiplying 9 times each entire side:

$$(9)\left(\frac{x^2}{9} - \frac{x}{3}\right) = (9)1 \rightarrow \frac{9x^2}{9} - \frac{9x}{3} = 9 \rightarrow x^2 - 3x = 9.$$

Ah, much better. Now, let's subtract 9 from both sides for standard form: $x^2 - 3x - 9 = 0$. Now we know that $a = 1$, $b = -3$, and $c = -9$. We'll stick these into the formula that Betty helps us remember:

$$x = \frac{-b \pm \sqrt{b^2 - 4ac}}{2a} \rightarrow x = \frac{-(-3) \pm \sqrt{(-3)^2 - 4(1)(-9)}}{2(1)}$$

$$\rightarrow x = \frac{3 \pm \sqrt{9 + 36}}{2} \rightarrow x = \frac{3 \pm \sqrt{45}}{2}.$$ At this point, we notice the factor of 9 lurking inside the radical sign, so:

$$x = \frac{3 \pm \sqrt{9 \cdot 5}}{2} \rightarrow x = \frac{3 \pm 3\sqrt{5}}{2}.$$ There's nothing left to simplify!

Answer: $x = \dfrac{3 - 3\sqrt{5}}{2}, \dfrac{3 + 3\sqrt{5}}{2}$

2. $x^2 + 2x - 5 = 0$

3. $x^2 - 7 = 4x$

4. $\frac{x^2}{5} + \frac{2x}{5} + \frac{1}{5} = 0$

5. $8x^2 + 2x = 3$

6. $x^2 + 12x + 9 = 0$

(Answers on p. 412)

Quadratic Equations with No Solution?

Some quadratic equations *have no real solutions*, meaning that no real numbers* can satisfy them. Sound strange to you? Well, frankly, it is a little strange—especially because, as you'll see in Algebra II, these same equations actually have *imaginary* solutions.

It's not hard to see why $x^2 = -4$ has no real solution. I mean, what could x be? We can't square a real number and get something negative, after all.

. . . .

* See p. 399 for the definition of *real numbers*.

But how about something like this: $x^2 + 2x + 3 = 0$? It's not so obvious whether or not it has real solutions. It even looks like it could be factored. But it can't—just try! Sure, we could plug everything into the quadratic equation, but look what happens:

So, $a = 1$, $b = 2$, and $c = 3$, right? Plugging into $x = \dfrac{-b \pm \sqrt{b^2 - 4ac}}{2a}$, we get:

$$x = \frac{-(2) \pm \sqrt{(2)^2 - 4(1)(3)}}{2(1)} \rightarrow x = \frac{-2 \pm \sqrt{4 - 12}}{2} \rightarrow x = \frac{-2 \pm \sqrt{-8}}{2}$$

Wait, what? How can we have a negative sign inside the radical $\sqrt{-8}$? That's not a real number! Well that's true, and we've just learned that this equation has *no real solutions*.

If only we'd known that the stuff under the square root sign was negative *ahead* of time, we could have saved a bunch of work. So why not just take a look at $\sqrt{b^2 - 4ac}$ first, and immediately know where we stand? I'm glad you asked.

Shortcut Alert

The discriminant is the stuff underneath the square root sign in the quadratic formula, and we can use it to find out *how many real solutions* a quadratic equation will have. So, once you've written your equation in standard form, $ax^2 + bx + c = 0$, just figure out what a, b, and c are, and then evaluate the **discriminant**:

$$b^2 - 4ac$$

If $b^2 - 4ac < 0$, then the equation has <u>no</u> real solutions.

If $b^2 - 4ac > 0$, then the equation has <u>two</u> real solutions.

If $b^2 - 4ac = 0$, then the equation has <u>one</u> real solution (a "double root").

It makes sense that if we get a negative sign under the square root sign, there can't be any real solutions, right? We've also seen plenty of examples with positive discriminants giving us two solutions.

And how about the last case, where the discriminant equals zero? This also makes sense, because if the square root becomes zero, that means we end up with $\pm\sqrt{0}$ on top of the fraction. The $\pm$ would then disappear, and there wouldn't be two solutions anymore. (We'll do an example of this in problem 1 below.)

Look, discrimination is not cool...unless we're talking about quadratic equations. Then we *should* discriminate against the ones that have no real solutions and save ourselves a bunch of time!

Doing the Math

For each equation, use the discriminant to find how many real solutions it has, and if there are any, then use the method of your choice to solve it. I'll do the first one for you.

1. $2x^2 - 4x = -2$

<u>Working out the solution</u>: Before we do anything, let's divide both sides of the equation by 2, so we get $\frac{2x^2 - 4x}{2} = \frac{-2}{2} \rightarrow x^2 - 2x = -1$. Then, adding 1 to both sides, we get the standard form: $x^2 - 2x + 1 = 0$. Okay, so $a = 1$, $b = -2$, and $c = 1$. We'll plug these into the discriminant: $b^2 - 4ac = (-2)^2 - 4(1)(1) = 4 - 4 = 0$. According to the shortcut on p. 393, we'll have only **one solution**. Let's find it! We already know that the radical sign becomes zero, so we can just do $x = \frac{-b \pm \sqrt{b^2 - 4ac}}{2a} \rightarrow x = \frac{-(-2) \pm \sqrt{0}}{2(1)} \rightarrow$ $x = \frac{2}{2} \rightarrow x = 1$. Let's check it by plugging $x = 1$ into the original equation: $2x^2 - 4x = -2$ becomes $2(1)^2 - 4(1) \stackrel{?}{=} -2 \rightarrow$ $2 - 4 \stackrel{?}{=} -2 \rightarrow -2 = -2$ Yep! So $x = 1$ is the solution.*

Answer: There is one real solution, $x = 1$.

• • • • • • • • • • •

* This also would have factored as $(x - 1)(x - 1) = 0$.

2. $x^2 - 3 = 2x$

3. $x^2 + 25 = 0$

4. $x^2 + 25 = 10x$

5. $-x^2 + 4x = 6$ *(Watch the negatives!)*

(Answers on p. 412)

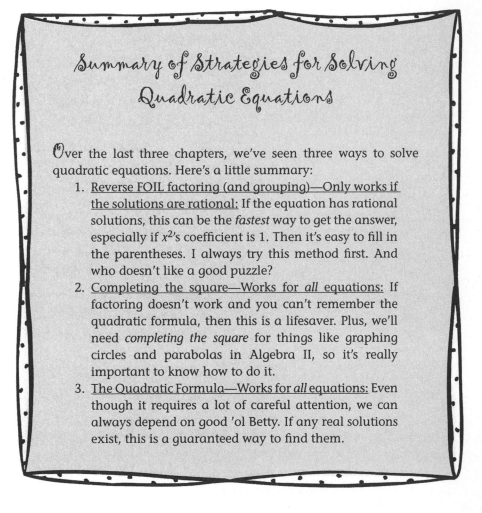

Summary of Strategies for Solving Quadratic Equations

Over the last three chapters, we've seen three ways to solve quadratic equations. Here's a little summary:

1. <u>Reverse FOIL factoring (and grouping)—Only works if the solutions are rational:</u> If the equation has rational solutions, this can be the *fastest* way to get the answer, especially if x^2's coefficient is 1. Then it's easy to fill in the parentheses. I always try this method first. And who doesn't like a good puzzle?

2. <u>Completing the square—Works for *all* equations:</u> If factoring doesn't work and you can't remember the quadratic formula, then this is a lifesaver. Plus, we'll need *completing the square* for things like graphing circles and parabolas in Algebra II, so it's really important to know how to do it.

3. <u>The Quadratic Formula—Works for *all* equations:</u> Even though it requires a lot of careful attention, we can always depend on good 'ol Betty. If any real solutions exist, this is a guaranteed way to find them.

What's the Deal?

Completing the Square and the Quadratic Formula

Believe it or not, the quadratic formula <u>is what you end up with</u> if you complete the square with $ax^2 + bx + c = 0$. Yup, just follow the "solving quadratic equations by completing the square" steps on pp. 374–75, and if you do everything correctly, you'll actually end up with

$x = \dfrac{-b \pm \sqrt{b^2 - 4ac}}{2a}$. Check out danicamckellar.com/hotx for a PDF of it. Or for a *really* big challenge, try it on your own! *(Hint: Start by dividing everything by a, complete the square, and then take the square roots of both sides. Good luck!)*

Takeaway Tips

 Memorize the quadratic formula: $x = \dfrac{-b \pm \sqrt{b^2 - 4ac}}{2a}$.
Say it out loud: "*x* equals negative *b*, plus or minus the square root of *b* squared minus 4*ac*, ALL over 2*a*," and remember Betty and the mall!

 Avoid careless mistakes by writing your equation in standard form, $ax^2 + bx + c = 0$; writing out the values for *a*, *b*, and *c*; and using plenty of parentheses along the way.

The discriminant, $b^2 - 4ac$, tells us how many real solutions our equation will have.

A Final Word

Ever experienced a brain haze? This can happen any time we are faced with something big and time-consuming that feels like an obstacle. It could be that you need to reorganize an entire drawer of beads that just spilled to the floor, or you have a math problem that you just don't know how to start. It's like the brain goes cloudy and wants to reject it: "This is way too complicated—never mind."

But these are the moments when you earn the right to become a truly successful person. Whenever you're confronted with a hard problem that makes your brain start to cloud up, practice thinking, "Wow, this looks challenging. It's a good thing I have *me* on my side." And practice <u>feeling good</u> as you say it. This will help relax your mind and allow the clouds to go away. Above all, practice seeing yourself as a problem solver. There are no "other" people who are more equipped than you to succeed in algebra or anything else you set your mind to. It's all about your attitude.

Read this book again and again. Reviewing math is never a waste of time. Not only does it keep your brain sharp, but all of these topics will come up again in future math classes. Plus, when you reread a section, you'll be surprised at the new things you pick up! All the time you dedicate to strengthening your mind and attitude will serve you not only in math class, but also in your career and in all areas of your life. I am so incredibly proud of you!

Danica

Appendix

$\mathcal{H}$ere are some definitions and number properties from pre-algebra and algebra for your reference. This is going to be dense; to read where I actually *teach* some of this stuff, see the various sections of *Kiss My Math* that I'll refer to throughout the Appendix.

Sets of Numbers

Set just means "collection." So a set of numbers is just the collection of a particular type of number. Here are some good *sets* to know:

Sets of Numbers

Counting Numbers: This is the set of all positive numbers we use for counting, starting with 1: {**1, 2, 3, 4, 5 . . .** }. This set does not include any fractions or decimals of any kind.

Whole Numbers: This is just the counting numbers with zero added. I know, I know—why the "whole" new name? Who knows, but here's the set: {**0, 1, 2, 3, 4 . . .** }.

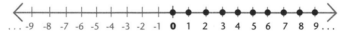

Integers: This is the set of all whole numbers and their opposites.* So it extends in both directions forever, positive and negative. Also, it does not include any fractions or decimals. Check it out: {**. . . –5, –4, –3, –2, –1, 0, 1, 2, 3, 4, 5, . . .** }. On a graph, this set of numbers is represented by dots at each number. Imagine them going on forever in both directions.

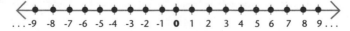

. . . .

* To review integers (or mint-egers!), check out Chapter 1 in *Kiss My Math*.

Rational Numbers: This is the set of all logical, level-headed numbers. Just kidding. It's the set of all values that can be expressed as a *ratio* of two integers; in other words, expressed as a fraction of two integers. All fractions, terminating decimals, and repeating decimals* are rational numbers, because they can be expressed as fractions of one whole number divided by another. Examples of rational numbers are 0.5, –1, $\frac{163}{1}$, $0.\overline{8}$, and $-\frac{4}{99}$.

Irrational Numbers: This is not the set of crazy numbers: It's the set of all values that *cannot* be expressed as a ratio of integers; for example, π, $\sqrt{2}$, 0.12345678910111213 . . .

Real Numbers: This is the set of all rational *and* irrational numbers—every value on the number line!

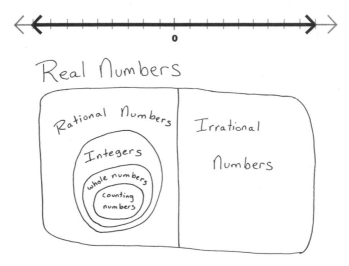

Some Number Properties

The associative, commutative, and distributive properties are actually ways around the whole PEMDAS thing. For the distributive property, see pp. 30–31 in this book.

Associative Property for Addition: For all numbers *a*, *b*, and *c*:

$$(a + b) + c = a + (b + c)$$

This says that if the *only operation is addition,* you can move around the parentheses; for example, (3 + 2) + 1 = 3 + (2 + 1). After all, they both equal 6.

· · · · · · · · · ·

* For a review of repeating decimals, check out Chapter 12 in *Math Doesn't Suck.* To see the proof of $0.\overline{9} = 1$, check out danicamckellar.com/hotx.

Associative Property for Multiplication: For all numbers *a, b,* and *c:*

$$(a \times b) \times c = a \times (b \times c)$$

This says that as long as *the only operation is multiplication,* then you can move around the parentheses; for example: $(6 \times 4) \times 2 = 6 \times (4 \times 2)$. They both equal 48, after all! Remember: The placement of parentheses matters because in the order of operations, we always do what's inside the parentheses first.

The Commutative Property for Addition: For all numbers *a* and *b,*

$$a + b = b + a$$

For example, $-13 + 5 = 5 + (-13)$. They both equal -8, after all!

The Commutative Property for Multiplication: For all numbers *a* and *b,*

$$a \times b = b \times a$$

For example, 3×4 equals 4×3. They both equal 12, after all!

For a full review of how the associative and commutative properties work and to learn some super-easy ways to remember them, check out Chapter 2 in *Kiss My Math.*

PEMDAS and PANDAS

Ah, pandas . . . aren't they cute? And they have really big appetites. I've heard that pandas like to eat dumplings with mustard, and then for dessert, they have apples with spice—like cinnamon or nutmeg. Yum!*

The *Order of Operations* is the order in which we must simplify math expressions:

PANDAS	**P**arentheses
EAT	**E**xponents
MUSTARD ON **D**UMPLINGS	**M**ultiplication & **D**ivision (whichever comes first)
AND	
APPLES WITH **S**PICE	**A**ddition & **S**ubtraction (whichever comes first)

For more explanation and to review how this works, see pp. 104–5 in *Math Doesn't Suck* or pp. 21–22 in *Kiss My Math.* And remember: **P**andas **E**at **M**ustard on **D**umplings and **A**pples with **S**pice!

. . . .

* The truth? Pandas eat mostly bamboo. Just go with me here, kay?

Squares and Square Roots

Here's a little table to help you with squares and square roots. It's not a bad idea to give this a read now and then; it's helpful to be familiar with them. . . .

Squares of Integers from 1 to 16	Square Roots of Integers from 1 to 16
You'll recognize most of these from your multiplication facts!	*Exact square roots are bold; all others are approximations to irrational values.*
Squares	**Square Roots**
$1^2 = 1$	$\sqrt{\mathbf{1}} = \mathbf{1}$
$2^2 = 4$	$\sqrt{2} \approx 1.414$
$3^2 = 9$	$\sqrt{3} \approx 1.732$
$4^2 = 16$	$\sqrt{\mathbf{4}} = \mathbf{2}$
$5^2 = 25$	$\sqrt{5} \approx 2.236$
$6^2 = 36$	$\sqrt{6} \approx 2.449$
$7^2 = 49$	$\sqrt{7} \approx 2.646$
$8^2 = 64$	$\sqrt{8} \approx 2.828$
$9^2 = 81$	$\sqrt{\mathbf{9}} = \mathbf{3}$
$10^2 = 100$	$\sqrt{10} \approx 3.162$
$11^2 = 121$	$\sqrt{11} \approx 3.317$
$12^2 = 144$	$\sqrt{12} \approx 3.464$
$13^2 = 169$	$\sqrt{13} \approx 3.606$
$14^2 = 196$	$\sqrt{14} \approx 3.742$
$15^2 = 225$	$\sqrt{15} \approx 3.873$
$16^2 = 256$	$\sqrt{\mathbf{16}} = \mathbf{4}$

Answer Key

For the fully worked out solutions, visit the "Solution Guides" page at danicamckellar.com/hotx.

Chapter 1, p. 8

2. A

3. D

4. C

5. B

Chapter 1, pp. 12–13

2a. $n + (n + 1) = -11$

2b. $-6, -5$

3a. $n + (n + 1) + (n + 2) = 42$

3b. 13, 14, 15

4a. $n + (n + 2) + (n + 4) = -3$

4b. $-3, -1, 1$

Chapter 2, pp. 20–21

2. ab

3. $14xy^2$

4. 5

5. $2a^2b$

Chapter 2, p. 26

2. GCF: jk; LCM: $7jk^2$

3. GCF: $11mn$; LCM: $66m^2n^2$

4. GCF: b; LCM: $36a^2b$

Chapter 3, pp. 34–35

2. $2b(a + 5b)$

3. $cd(6c - 7d)$

4. $9x$

Chapter 3, pp. 37–38

2. $5b(3a - 4b)$

3. $6bcd(2b - 3)$

4. $6bd(2bc - 3c + 4)$

5. $xy(xy + x + y - 1)$

Chapter 4, p. 44

2. $\dfrac{ac}{2b}$

3. $\dfrac{2b}{ac}$

4. $\dfrac{3e^2f}{20}$

5. $\dfrac{10}{n}$

Chapter 4, pp. 49–50

2. $\dfrac{3a - 1}{2 + c}$

3. $3y + 1$

4. $\dfrac{11d}{7}$

5. -4

Chapter 4, p. 54

2. $\dfrac{4}{y}$

3. 1

4. $\dfrac{2(x + 1)}{3}$ or $\dfrac{2x + 2}{3}$

Chapter 5, p. 59

2. $\dfrac{3x}{5}$

3. $\dfrac{3x}{y}$

4. $-\dfrac{a}{b}$ or $\dfrac{-a}{b}$

Chapter 5, p. 65

2. $\dfrac{7}{2a}$

3. $\dfrac{v + 6}{3v}$

4. $\dfrac{1}{n(1 - n)}$ or $\dfrac{1}{n - n^2}$

5. $\dfrac{xy - 1}{y}$

6. $\dfrac{2a^2 + 6b + 3c}{12ab}$

7. $\dfrac{8y - 1}{2y(y - 1)}$ or $\dfrac{8y - 1}{2y^2 - 2y}$

Chapter 5, pp. 68–69

2. $9 + x^2$

3. $\dfrac{cd}{c + d}$

4. 0

Chapter 6, pp. 80–81

2a. $y = 2 - 4x; y = 2, -6$

2b. $x = \dfrac{2 - y}{4}; x = \dfrac{1}{4}, -\dfrac{7}{4}$

3a. $y = \dfrac{x + 1}{2}; y = \dfrac{1}{2}, \dfrac{3}{2}$

3b. $x = 2y - 1; x = 1, 17$

4a. $y = 12x; y = 0, 24$

4b. $x = \dfrac{y}{12}; x = \dfrac{1}{12}, \dfrac{3}{4}$

Chapter 6, pp. 87–88

2. $x = \dfrac{c}{a + b}$

3. $x = \dfrac{14}{y + z - 4}$

4. $x = \dfrac{d + 5}{d}$

5. $x = 1$

Chapter 7, pp. 97–98

2. $-3, 9$

3. $-9, 3$

4. $0, 4$

5. $-8, -2$

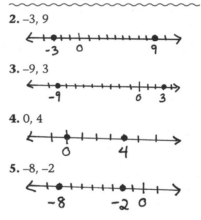

Chapter 8, p. 105

2. disjunction,

disjunction: $n \leq 3$ or $n > 6$

3. no solution

4. conjunction

conjunction: $y > -4$ and $y \leq -1$
$-4 < y \leq -1$

5. conjunction

conjunction: $w \geq 5$ and $w < 8$
$5 \leq w < 8$

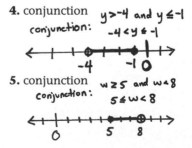

Chapter 8, p. 111

2. disjunction,

$$y \leq -1 \quad \text{or} \quad y \geq 3$$

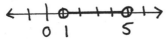

3. conjunction

$$1 < a < 5$$

4. conjunction

$$-1 \leq x \leq 1$$

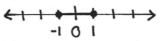

5. disjunction

$$n < -2 \quad \text{or} \quad n > 1$$

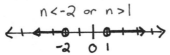

6. no solution

Chapter 9, p. 122

2. function; domain: {0, 2, 3}; range: {-3, 0, 2}

3. relation; domain: {-1}; range {1, 3, 9, 10}

4. function; domain: {-7, 0, 3, 5}; range: {-4, 8}

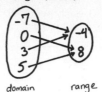

5. function; domain: {2, 3}; range: {6}

Chapter 9, pp. 127–28

2. function; domain: $x \geq 0$; range: $y \geq 0$.

3. function; domain: $-4 < x \leq 1$; range: $-2 \leq y < 2$.

4. relation; domain: $x > 0$; range: all real numbers except $y = 2$.

5. function; domain: $x > -1$; range: $y < 1$.

Chapter 10, pp. 137–38

2. a. yes; **b.** $x + y = 4$; **c.** $y = -x + 4$; **d.** slope = -1, y-intercept = 4, x-intercept = 4; **e.** shown with points (0, 4), (2, 2), (4, 0)

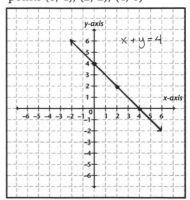

3. a. yes; **b.** $2x - 5y = 0$; **c.** $y = \frac{2}{5}x$;
d. slope $= \frac{2}{5}$, y-intercept $= 0$,
x-intercept $= 0$, **e.** shown with
points $(-5, -2)$, $(0, 0)$, $(5, 2)$

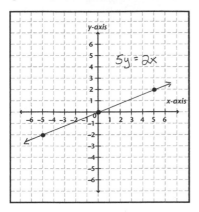

4. a. no; **b.** $4x - 2y = -7$;
c. $y = 2x + \frac{7}{2}$; **d.** slope $= 2$,
y-intercept $= \frac{7}{2}$, x-intercept $= -\frac{7}{4}$;
e. shown with points $\left(-\frac{7}{4}, 0\right)$,
$\left(0, \frac{7}{2}\right)$, $\left(1, \frac{11}{2}\right)$

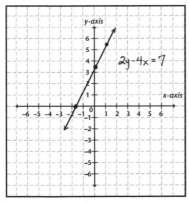

5. a. yes; **b.** $3x + y = 6$;
c. $y = -3x + 6$; **d.** slope $= -3$,
y-intercept $= 6$, x-intercept $= 2$;
e. shown with points $(1, 3)$,
$(2, 0)$, $(3, -3)$
[See top of next column for
graph.]

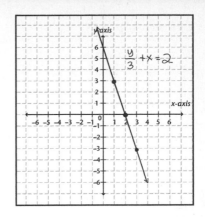

Chapter 10, p. 144

2. shown with points $(-1, -2)$,
$(0, 1)$, $(1, 4)$:

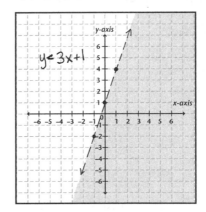

3. shown with points $(-2, 0)$,
$(-1, -2)$, $(0, -4)$:

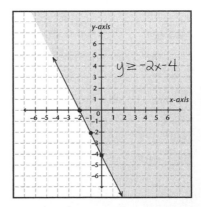

4. shown with points (5, –2), (5, 0), (5, 2):

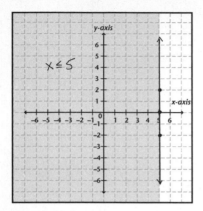

5. shown with points (0, 2), (3, 0), (6, –2)

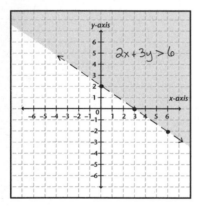

Chapter 11, p. 149

2. $y = 2x - 3$

3. $y = 3x - 1$

4. $y = -x$

5. $y = \frac{1}{4}x + 4$

Chapter 11, pp. 155–56

2. $y = 3x$

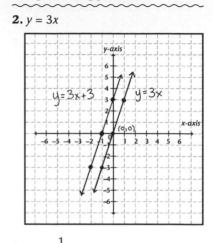

3. $y = -\frac{1}{3}x$

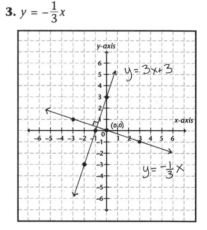

4. $x = 2$

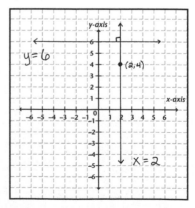

5. $y = -\dfrac{2}{3}x + 3$

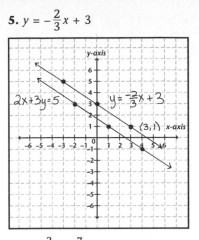

6. $y = \dfrac{3}{2}x - \dfrac{7}{2}$

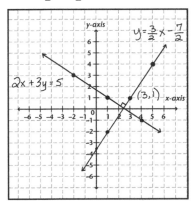

7. $y = 4x - 3$

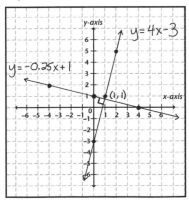

Chapter 11, pp. 160–61

2. $m = 1;\ y = x + 2$

3. $m = -\dfrac{1}{2};\ y = -\dfrac{1}{2}x + 1$

4. $m = \dfrac{2}{3};\ y = \dfrac{2}{3}x + 5$

5. $m = -\dfrac{7}{2};\ y = -\dfrac{7}{2}x + 5$

Chapter 12, p. 168

2. $x = 2,\ y = 6$

3. $x = -1,\ y = 5$

4. $x = 2,\ y = \dfrac{1}{4}$

Chapter 12, p. 174

2. $x = 2,\ y = -3$

3. $x = -4,\ y = \dfrac{24}{5}$

4. $x = 1,\ y = 1$

5. $x = -4,\ y = 3$

Chapter 12, pp. 180–81

2. consistent & independent;
solution: (0, 4)

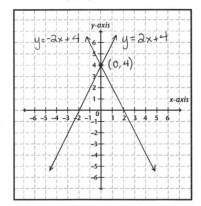

3. inconsistent; no solution

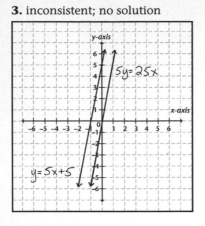

4. consistent & dependent;
solution: entire line

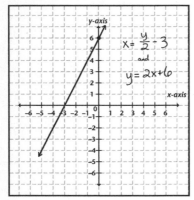

5. consistent & independent;

solution: $\left(\dfrac{12}{7}, \dfrac{29}{7}\right)$

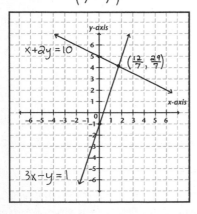

6. consistent & independent;
solution: (4, 3)

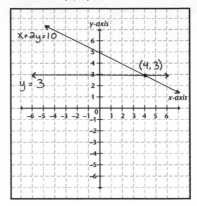

Chapter 13, pp. 193–95

2. 10 miles/hour

3. $\dfrac{1}{2}$ hour

4. 2 km/hour

5. 1,200 miles from NY

6. 3 hours

Chapter 14, pp. 204–6

2. 20 hours

3. 2 hours, 24 minutes

4. 1 hour, 12 minutes

5a. 2 invitations/hour

5b. 1 hour, 30 minutes

6. 24 minutes

Chapter 15, pp. 213–14

2. $49.50

3. $60.50

4.a $4; b. $7

5. $80

6. $468

Chapter 15, p. 220

2. $2.50

3. $1000

4. 24 years

Chapter 16, pp. 230–31

2. 4 nickels

3. 12 pairs high heels

4. 5 lbs jelly beans

5. 9 dimes

6. 2 chores

Chapter 16, pp. 237–38

2. $1\frac{1}{3}$ gal Brand A; $\frac{2}{3}$ gal Brand B

3. 8 gal pure glitter

4. 5 oz C.Charlie; 15 oz R.Rosy

5. 20 gal saltwater

6. 24 grams pure gold

Chapter 17, p. 245

2. x^6y^7

3. $25y^6$

4. $4x^3yz$

Chapter 17, p. 250

2. $16a^4b^8$

3. $-16a^4b^8$

4. $16a^4b^8$

5. $-\dfrac{x^9y^3}{125}$

6. $xy^{14}z^4$

Chapter 17, p. 254

2. $\dfrac{b^2}{a^2}$

3. $\dfrac{x}{8w^2}$

4. $3y$

Chapter 18, pp. 261–62

2. 16

3. $\dfrac{2e^3}{f}$

4. $4x$

5. $1 + \dfrac{1}{25z^2}$

6. $\dfrac{a^2}{8}$

7. $\dfrac{1}{c^3} + 1$

Chapter 18, pp. 267–68

2. $\dfrac{7}{6}$

3. $\dfrac{64}{b^{11}}$

4. 1

5. $\dfrac{w^4(1-y)}{2+x}$

6. $\dfrac{q^3}{81}$

Chapter 19, p. 279

2. 5

3. 6.09

4. not a real number

5. –0.9

Chapter 19, pp. 284–85

2. $5\sqrt{2}$

3. $2\sqrt{21}$

4. $6\sqrt{3} + 15$

5. $18 + 2\sqrt{17}$

Chapter 19, p. 287

2. 6

3. $40\sqrt{6}$

4. $70\sqrt{3}$

Chapter 19, p. 290

2. $7\sqrt{5}$

3. $7\sqrt{5}$

4. $\sqrt{5}$

5. $-\sqrt{5}$

6. $-2\sqrt{5}$

Chapter 20, p. 299

2. a. 3 terms; **b.** $6wz^2 - 1$;
   c. degree = 3; 2 terms; binomial

3. a. 4 terms; **b.** 8; **c.** degree = 0; 1
   term; monomial

4. a. 5 terms; **b.** $-2d^2e^3 + e^4 + 6$;
   c. degree = 5; 3 terms; trinomial

Chapter 20, pp. 300–301

2. $a^4b^4(2a + 6b + 9ac)$

3. $3z^3(z^5 - 4y^2z^2 + 6y^3z - 2)$

4. $7gh^6k(2k - 6g^3h + 7k^3)$

Chapter 20, pp. 303–4

2. $-3b^2$

3. $g^2h + gh^2 - 2gh$

4. $y^3 - 1$

5. $x^6 - 5x^5 + 2x^4 + 3x^3 - 11$

Chapter 21, p. 313

2. $2x^2 - 5x - 3$

3. $zw^2 + 2w^2 - 3z - 6$

4. $a^2 - b^2$

Chapter 21, p. 315

2. $x^2 - 2x + 1$

3. $9y^2 - 12y + 4$

4. $a^2 + 2ab + b^2$

5. $a^2 - 2ab + b^2$

Chapter 21, p. 317

2. $x^2 - xy + 2y - 4$

3. $27 + y^3$

4. $a^3 + b^3$

5. $2h^4 + h^2 - 2g^2 + 5g - 3$

Chapter 22, p. 326

2. 2 & 1   **6.** 9 & –8

3. –12 & 1   **7.** –3 & 2

4. 3 & –2   **8.** –5 & –4

5. –8 & 1   **9.** –6 & –3

Chapter 22, pp. 329–30

2. $(x + 1)(x + 7)$ **6.** $(y + 3)(y - 5)$

3. $(w - 3)(w + 4)$ **7.** $(x - 2)(x - 9)$

4. $(h - 1)(h - 11)$ **8.** $(g - 2)(g + 9)$

5. $(x - 3)(x + 5)$ **9.** $(x - 3)(x - 6)$

Chapter 22, p. 336

2. $(2x + 1)(x + 1)$

3. $(7h + 1)(h + 2)$

4. $(2w + 1)(3w + 1)$

5. $(2x + 1)(x - 2)$

6. $(10m + 3)(m + 3)$

7. $-5(2y + 1)(y - 2)$

8. $(4g + 1)(g - 1)$

9. $(3x - 1)(x + 3)$

Chapter 22, p. 341

2. $(4x - 3)(3x + 2)$

3. $(2x + 3)(5x - 2)$

4. $(x - 2)(3x + 1)$

5. $(2x + 3)(4x - 5)$

Chapter 23, p. 345

2. $(x + 3)(x - 3)$

3. $2(x + 3)(x - 3)$

4. $(2w + 5)(2w - 5)$

5. $3(2w + 5)(2w - 5)$

6. $(3x - 2)^2$

7. $5x(y + 3)(y - 3)$

8. $(4h + 1)^2$

9. $3x(3x - 2)^2$

Chapter 24, p. 356

2. $x^2 + x - 2 = 0$; $x = -2, 1$

3. $y^2 + 4y + 4 = 0$; $y = -2$

4. $3t^2 - 13t + 4 = 0$; $t = \frac{1}{3}, 4$

5. $6x^2 - x - 15 = 0$; $x = -\frac{3}{2}, \frac{5}{3}$

Chapter 24, p. 360

2. $x = -7, 1$

3. $x = -1, 7$

4. $w = -\frac{1}{2}, 1$

5. $y = -1, 6$

6. $x = -5, 4$

7. $h = -1, \frac{1}{4}$

Chapter 24, pp. 363–64

2. $x^2 - 4x - 5 = 0$

3. $x^2 + 4x - 5 = 0$

4. $x^2 - 6x + 5 = 0$

5. $x^2 + 6x + 5 = 0$

6. $15x^2 - 7x - 2 = 0$

7. It's up to you!

Chapter 25, p. 371

2. $y = -3, 5$; $y^2 - 2y - 15 = 0$

3. $t = -5, -1$;

$t^2 + 6t + 5 = 0$

4. $x = 3 - \sqrt{5}, 3 + \sqrt{5}$;

$x^2 - 6x + 4 = 0$

Chapter 25, p. 378

2. $(x + 4)^2 = 17$

3. $(x + 1)^2 = 3$

4. $(x - 3)^2 = 12$

5. $\left(x + \frac{1}{2}\right)^2 = \frac{9}{4}$

Chapter 25, p. 380

2. $d^2 = 25$; $x = -9, -1$

3. $d^2 = 1$;

$x = -1 - \sqrt{5}, -1 + \sqrt{5}$

4. $d^2 = 4$;

$x = 2 - \sqrt{6}, 2 + \sqrt{6}$

5. $d^2 = \frac{9}{4}$; $x = -4, 1$

Chapter 26, pp. 391–92

2. $x = -1 - \sqrt{6}, \; -1 + \sqrt{6}$

3. $x = 2 - \sqrt{11}, \; 2 + \sqrt{11}$

4. $x = -1$, a double root

5. $x = -\dfrac{3}{4}, \dfrac{1}{2}$

6. $x = -6 - 3\sqrt{3}, \; -6 + 3\sqrt{3}$

Chapter 26, pp. 394–95

2. Two real solutions: $x = -1, 3$

3. No real solutions

4. One real solution: $x = 5$

5. No real solutions

Index

Huck, Sarah, 306–308
Hypotenuse, 280–281

Inconsistent equations, 176–182
Independent equations, 177–182
Inequalities (*see also* Linear
    Inequalities), 100–105
    with absolute value, 106–113
Integers
    consecutive, 10–13
    defined, 398
Inverse operations, 77–78, 275
Invisible exponents, 20, 243, 247,
    253–254, 264, 295, 304
Irrational numbers, 282, 286, 288,
    364, 379, 388*n*, 391, 399, 401
Isolating *x*, 78–79

Jobs per time, 198–205
Journal writing, 21, 75, 122–123,
    256, 273, 282–283, 287, 368

Labeling, 9–13, 70, 186–187, 201,
    218, 226–228, 232–233, 238,
    361
Landow, Robyn, 72, 348
LCD (lowest common
    denominator), 61–64, 84
LCM (least common multiple),
    22–27, 61
Least common multiple (*see* LCM)
Like monomials, 296, 302
Like terms, combining, 6, 35, 243,
    297, 299, 302–304, 314, 320
Linear equations (*see also* Lines)
    "Adding/Subtracting Twins"
    (elimination) method and,
    169–174, 181–182
    graphing systems of, 175–181
    solving systems of, 164–175,
    178–182
    substitution method and,
    165–168, 181*n*, 182
Linear functions (*see also* Linear
    equations), 130–131
Linear inequalities, graphing,
    141–145
Lines
    horizontal/vertical, 124*n*, 133*n*,
    139, 154

parallel/perpendicular, 150–156,
    161, 176–177, 180, 182
    slope-intercept form of, 131–132,
    134–138, 145, 152–155, 160,
    161
    standard form of, 132–134,
    137–138
Lowest common denominator
    (*see* LCD)
Lowest common multiple (*see* LCM)
Lowney, Anne, 72, 348

Mapping diagrams, 117–122
Mathathon, 223*n*, 418
McKellar, Crystal, 146
Mixture problems, 224–238
Money management, 10, 82,
    208–221, 223, 321, 384
Monomials, 294–299, 302, 304
Motion problems, 184–196, 231
Multiples (*see also* LCM), 21–22
Multiplying
    decimals, 86–87, 212, 220,
    235–237
    fractions and reciprocals, 4–7,
    42–44, 51–53
    polynomials, 309–320
    product of factors and, 15–16, 31
    radical expressions, 285–290

Negative exponents, 258–270
Negative numbers
    brushing up on, 12*n*, 48*n*
    streamlining, 4–6, 302–303,
    311–313, 315, 320
    square roots and, 276, 279, 370,
    381–382, 393
Negative rates, 206, 230, 231
Negative reciprocals, 151–155, 161
Number lines, 97–98, 100–113
Numerical coefficient
    (*see* Coefficients)

Ordered pairs, 118–124, 132,
    135–139, 142–145, 148–161

Parallel lines, 150, 152, 153, 156,
    161, 176–177, 180, 182
Parry, Courtney, 162–163

About the Author

Well known for her roles as Winnie Cooper on *The Wonder Years* and Elsie Snuffin on *The West Wing*, Danica McKellar is an internationally recognized mathematician and advocate for math education and is also a two-time *New York Times* bestselling author.

Upon the release of her groundbreaking bestseller *Math Doesn't Suck*, McKellar made headlines and was named "Person of the Week" by ABC World News with Charles Gibson. Her second book, *Kiss My Math: Showing Pre-Algebra Who's Boss,* followed suit, debuting at #4 on the *New York Times* bestseller list. McKellar has been honored in Britain's esteemed *Journal of Physics* and the *New York Times* for her prior work in mathematics, most notably for her role as coauthor of a mathematical physics theorem that now bears her name (the Chayes-McKellar-Winn Theorem).

A summa cum laude graduate of UCLA with a degree in Mathematics, McKellar's passion for promoting girls' math education has earned her multiple invitations to speak before Congress on the importance of women in math and science. Amidst her busy acting schedule, she continues to make math education a priority, as a featured guest and speaker at mathematics conferences nationwide.

McKellar is also spokesperson for the Math-A-Thon program at the St. Jude Children's Research Hospital, which raises millions of dollars every month both for cancer research and to provide free care for young cancer patients.

McKellar lives in Los Angeles, California, and this is her third book.